Relational Land-Based Science, Technology, Engineering, Arts and Mathematics (STEAM) Education

Bios-Mythois

Jennifer D. Adams
Series Editor

Vol. 1

Relational Land-Based Science, Technology, Engineering, Arts and Mathematics (STEAM) Education

Edited by Eun-Ji Amy Kim and Kori Czuy

PETER LANG

New York · Berlin · Bruxelles · Chennai · Lausanne · Oxford

Library of Congress Cataloging-in-Publication Control Number: 2023006902

Bibliographic information published by the Deutsche Nationalbibliothek.
The German National Library lists this publication in the German
National Bibliography; detailed bibliographic data is available
on the Internet at http://dnb.d-nb.de.

Cover design by Kori Czuy

ISSN 2836-3159 (print)
ISBN 9781636671680 (paperback)
ISBN 9781636672441 (hardback)
ISBN 9781636670799 (ebook)
ISBN 9781636670805 (epub)
DOI 10.3726/b20652

© 2024 Peter Lang Group AG, Lausanne
Published by Peter Lang Publishing Inc., New York, USA
info@peterlang.com – www.peterlang.com

This publication has been peer reviewed.

to earth mother and all she gives us
to elders past, present, emerging
to all those that are healing
hand to heart

Contents

List of Figures/Tables

Foreword

DWAYNE DONALD

In September 1874, Treaty Commissioners representing the British Crown and Queen Victoria traveled to Fort Qu'Appelle to negotiate the terms of a sacred promise to live in peace and friendship with Indigenous peoples of the region that came to be known as Treaty 4. Prior to this meeting, Indigenous leaders had learned that the Hudson's Bay Company had sold their lands to the Dominion of Canada without their consent or consultation. Thus, when the Treaty Commissioners sought to initiate negotiations, they were surprised to learn that the leaders declined to discuss the Treaty. Instead, a chief spokesman named *The Gambler* explained with the help of a translator that the assembled peoples felt that there was "something in the way" of their ability to discuss the terms of the Treaty in good faith (Morris, 2014, pp. 97–98). It took several days of discussion for the Queen's representatives to comprehend the concerns expressed by *The Gambler*. The people were questioning the sincerity of these Treaty negotiations because they knew that the Government of Canada had already made a side-deal with the Hudson's Bay Company for the purchase of their lands. The view expressed by *The Gambler* was that these side dealings undermined the integrity of the Government's Treaty intentions. Through the translator, *The Gambler* clearly articulated the view that the Hudson's Bay Company really only had the permission of the Indigenous peoples to conduct trade. They did not have the right to claim ownership over any land: "The Indians want the Company to keep at their post and nothing beyond. After that is signed they will talk about something else" (p. 110). Despite these misunderstandings, as well as notable disagreement among the various Indigenous groups in attendance, the terms of Treaty 4 were eventually ratified.

I begin this Foreword with consideration of this particular microcosmic historical interchange because it is directly connected to the colonial macrocosm that this book brings focus on.[1] The founding principle of colonialism is relationship denial and the centuries-long epistemological predominance of this principle has resulted in the creation of institutional and societal structures that perpetuate relationship denial in mostly subtle and unquestioned ways. In both literal and metaphorical ways, the colonial impetus has resulted in the imposition of a macro-level grid system on top of land and people that has divided the world according to very arbitrary racial, cultural, and epistemological categorizations. With regard to the particular micro-event example from Treaty 4, the years 1830–1870 were the golden age of the fur trade on the Canadian Prairies—a time when the Hudson's Bay Company prospered and managed a fur trade monopoly over an expanding commercial network that tied together diverse groups of people in co-dependent relationships. However, as the years went on and more newcomers moved into the region, the governing systems began to change and the previous co-dependent balance was lost. During the 1850s, the lifestyles of people in the Prairie region gradually underwent a "Great Transformation" from a communal-style economy toward a private property system based on emerging forms of liberal economics and industrial capitalism (O'Riordan, 2003). As the significance of the land, resources, and people of the region began to be measured according to these emerging understandings of the market, ownership, and private property, previous co-dependent relational balances were subverted and a very violent form of enclosure—colonial frontier logic—was imposed on the land and the people in the name of Progress.[2] This is what *The Gambler* expressed as being 'in the way' of the Treaty negotiations.

As the Editors and Authors of this book have thoughtfully noted, this colonial frontier logic has also got "in the way" of schooling practices in the sense that prevailing curricular and pedagogical approaches perpetuate relationship denial. Those approaches are reflective of the "Western code"—the Enlightenment-based knowledge system that is presented as possessing all the answers to any important questions that could be asked (Mignolo, 2011, xii). It is important to state that conceptions of knowledge and knowing derived from such techno-scientific understandings of the world have provided many benefits. However, belief in the veracity of those understandings becomes a form of violence when that way of understanding is upheld as the only way. Wynter (1995), for example, has argued that the Columbian landfall on Turtle Island instigated a centuries-long hegemonic process wherein a universalized model of the human being was imposed on people around the world. Citing Foucault's "figure of Man", and noting the epistemological complexes resulting from Enlightenment-based arrangements of knowledge and knowing, Wynter asserts that this particular advancement has served to "absolutize the behavioural norms encoded in our present culture-specific conception of being human, allowing it to be posited as if it were the universal of the human species" (Wynter, 1995, pp. 42–43, emphasis in original). Eventually, formal schooling became a primary means by which those with power could discipline the citizenry to conform to this model of the human being. As I see it, this has resulted in the predominance of curricular and pedagogical approaches that perpetuate these universalized behavioural norms by persistently presenting knowledge and knowing in Enlightenment-based forms. As Lowe (2015) observes, the current moment is so replete with these universalized assumptions of human knowing and being that it has become very difficult to imagine other knowledge systems or ways of being human (p. 175).

This struggle to imagine other knowledge systems or ways of being human is implicated in the deepest difficulties faced today in trying to live in less damaging, divisive and destructive ways. It is clear to me that institutional and societal acceptance of relationship denial as the natural cognitive habit of human beings undermines the ability to respond to challenges in meaningful ways and proceed differently. The lingering influences of colonial frontier logics continues to get "in the way" of possibilities for relational repair and renewal. Instead of seeking to provide a specific solution to these challenges—"just do as I say and all will be better"— I have learned that it is much more helpful to focus on the creation of conditions under which something life-giving and life-sustaining can emerge. By taking seriously the wisdom of relational understandings of the world, the fine contributions provided in this book serve to create such conditions.

NOTES

1 In making this point, I am inspired by this piece of wisdom from Cajete (2004): "We are, after all, a microcosm of the macrocosm" (p. 47). I was reminded of this insight after reading Wiseman, Lunney Borden, and Sylliboy's chapter offering in this book.

2 I choose to capitalize this term to denote its mythological prominence within settler colonial societies like Canada. This notion of Progress has grown out of the colonial experience and is predicated on the pursuit of unfettered economic growth and material prosperity stemming from faith in market capitalism. For more on this see Donald (2019) and Nisbet (1980).

REFERENCES

Cajete, G. (2004). Philosophy of native science. In A. Waters (Ed.), *American Indian thought: Philosophical essays* (pp. 45–57). Malden, MA: Blackwell.

Donald, D. (2019). Homo economicus and forgetful curriculum. In H. Tomlinson-Jahnke, S. Styres, S. Lille, & D. Zinga (Eds.), *Indigenous education: New directions in theory and practice* (pp. 103–125). University of Alberta Press.

Lowe, L. (2015). *The intimacies of four continents.* Duke University Press.

Mignolo, W. D. (2011). *The darker side of Western modernity: Global futures, decolonial options.* Duke University Press.

Morris, A. (2014). *The Treaties of Canada with the Indians of Manitoba and the North-West Territories: Including the negotiations on which they are based, and other information relating Thereto.* Cambridge University Press.

Nisbet, R. A. (1980). *History of the idea of progress.* Transaction.

O'Riordan, T. (2003). Straddling the "Great Transformation": The Hudson's Bay Company in Edmonton during the transition from the commons to private property, 1854–1882. *Prairie forum, 28*(1), 1–26.

Wynter, S. (1995). 1492: A new world view. In V. L. Hyatt & R. Nettleford (Eds.), *Race, discourse, and the origin of the Americas: A new world view* (pp. 5–57). Smithsonian.

CHAPTER ONE

Grounding Selves and Intentions

EUN-JI AMY KIM & KORI CZUY

We (Kori and Eun-Ji Amy) will begin this book by grounding self and our intentions by introducing who we are and how this book came to be. As we travel this academic journey, our mentors continuously teach us the importance of clearly laying out the purpose and relationality of our intellectual/academic labours. We are reminded of the advice of Eber Hampton (1995), to connect our feelings with our work,

> Feeling is connected to our intellect and we ignore, hide from, disguise, and suppress that feeling at our peril and at the peril of those around us.
> Emotionless, passionless, abstract, intellectual, academic research is a goddamn lie, it does not exist. It is a lie to ourselves and a lie to other people (p. 52).

In laying down our purpose, motivation, and relations in compiling diverse stories presented in this book. We tune into our feelings as much as into our intellectual work. Often, these purposes and motivations are found within our stories (Kovach, 2009). We are committed to connect our minds with our hearts. This chapter begins this process, where layers of different narrative accounts from Kori and Eun-Ji Amy weave amongst our collective voices.

Through this process, we acknowledge our own individual learning journeys and spirits as well as our collective journeys to move beyond the colonial frontier logics. We, as editors of this collection, continue to reflect on the collective yet diverse journeys of moving beyond the colonial frontier logics through a Land-based, interdisciplinary STE(A)M approach. We acknowledge that there are different positions and roles given when we engage in Indigenous and Western ways of thinking and coming to know. Below, you will hear our—Kori and Eun-Ji Amy's—own stories, perspectives, experiences, and journey of engaging with Land-based Interdisciplinary STE(A)M education.

The initial idea of this book project goes back to the year 2019. When we (Kori and Eun-Ji Amy) along with Jennifer Adams and Mindi Lee Meadow presented at a panel session at the annual conference of the Canadian Society for the Study of Education (CSSE). The four of us are researchers and educators in the field of STEM education in various capacities. We came together because of our shared passion and commitments to making the STEM education space more inclusive towards multiple ways of coming to know. Our title for the symposium was, "Moving beyond the colonial frontier logics in STE(A)M education: A dialogue between Indigenous peoples & allies on Turtle Island through critical-transdisciplinary heuristics." The purpose of the panel was to re-imagine the conference space/place differently, and re-shuffle the dynamics at play at the conference. Instead of following the suit of the linear, hierarchical "expert" sharing their research ideas to the audience, we wanted a place of co-generative inquiry and storytelling.

Although the CSSE conference occupied the same physical space, there was a clear division in the cognitive space amongst the participants. This conference was organized similar to many others we all have attended, where people and ideas were separated based on racial and cultural categorizations thereby creating different Special Interest Groups (SIGs). Even informal events were divided based on racial and academic cultural categorizations. We also noted that relations were/are conceptualised as a simple binary distinction— Indigenous/Immigrants (settlers), colonized/colonizer. First of all, such a binary view does not acknowledge the diversity that exists within Indigenous communities. It promotes the pan-Indigenous approach, which leads to the essentialization of the diverse stories, histories, and relationality each community and Nation has with the Land and each other. It also promotes an essentialistic view of diverse populations of settlers and the diverse stories of their ancestors in relation with/on the land. The complex relations between diverse peoples and the stories of *how* they come to be on the land gets distilled and simplified by the construction of the binary Indigenous/Settlers (Haig-Brown, 2012).

Here, we turn to the notion of *colonial frontier logics* as a mechanism that operates by creating such a binary view of relations between peoples and ideas. Amiskwaciwiyiniwak and Papaschase Cree scholar, Dwayne Donald (2011) explains that,

Colonial frontier logics are those epistemological assumptions and presuppositions, derived from the colonial project of dividing the world according to racial and cultural categorizations, which serve to naturalize assumed divides and thus contribute to their social and institutional perpetuation (p. 92).

Colonial frontier logics, coupled with neoliberal globalization have played a significant role in constructing current global narratives in STEM curricula. These narratives are driven by neoliberal enclosure, where Western modes of thoughts and knowledges are being promoted through standardized testing systems such as PISA (Adams, Das, & Kim, 2020). Colonial frontier logics continue to divide knowledge based on racial and cultural (disciplinary cultures as well as ethnic cultures) categorizations. Such categorizations limit the relationship-based approach in teaching and learning. Meanwhile, we are currently witnessing a surge of a movement in Canadian teacher education institutions and K-12 education systems to develop land-based courses that include Indigenous ways of coming to know particularly after the publication of the *Accords on Indigenous Education* (Association of Canadian Deans of Education, 2010), calls to actions put forth by the Truth and Reconciliation Commission (TRC, 2015), and the 2021 Alberta Draft Curriculum[1] (of which I [Kori] cannot, will not get into here; primarily out of anger, but realistically, it would be a book unto itself).

For the 2019 CSSE panel symposium, we wanted diverse stories to come together within a space created specifically for co-generative inquiry. Particularly, we wanted to offer a space within the conference venues for Indigenous peoples and settlers to come together to share and make meaning on how educators and researchers, especially in the field of STEM, can bring together diverse knowledge systems and perspectives. We invited Dwayne Donald as our "discussant" and offered him a tobacco for sharing his knowledge and time with us. He shared the importance of Aboriginal and Settler peoples coming together to share stories while honouring the original stories told from the Land by the traditional custodians of the Land.

We wanted to continue to share and learn from one another. At the same time, we as facilitators of this sharing acknowledge the multiple realities, stances, and experiences individual participants may go through in their own journey in life. We put forth a call of chapters, first inviting all the participants who came to our panel to share with us their own journey of re-imagining and re-living the STEM curricula space with the Land.

While we appreciate the extensive labour and efforts placed in so-called *Indigenization* or *Reconciliation*, we remain critical. We ask, in what ways could we—diverse stakeholders in the field of education—do real work in moving beyond the colonial frontier logics in STEAM education? Also, how can the conceptualization/relationship with *Land* enhance the settler-Indigenous relations in the STEAM curriculum space?

BOOK LAYOUT

The cover artwork was created by Kori (explained further in the "A in STEAM" section) to represent our own editing processes of working with and bringing together diverse stories, ideas, and concepts within this book. The resulting artwork weaves together the relationships between the writers, researchers, students, alongside their learnings, experiences, and teachings.

The chapters in this book explore these questions through multiple ways of coming to know STEAM as expressed through art and story. As facilitators of these conversations, we focused on finding connections and emerging relations amongst diverse stories and concepts. Although the topic of STEAM weaves throughout the book, we decided to highlight some themes that arose from the authors stories. As such, we curate these chapters in four different themes: *Forts, Land-based, Remembering & Relationality,* and *A in STEAM.*

Forts

Dwayne used the Fort as an analogy to further illustrate the concept of colonial frontier logics. The Fort inherently represents the history of Aboriginal and Settler relations. Dwayne (2009) describes the Fort as, "a colonial artefact" representing "a particular four-cornered erosion of imperial geography that has been transplanted on lands perceived as empty and unused" (p. 3). Indeed, with the understanding of the land as empty land—*terra nullius*—many colonial projects in Commonwealth countries including Australia and Canada made it possible for settlers to claim the land as their own and began the nation building process seemingly "from scratch." Settlers started building livelihood within the fort, with the walls used to "protect" this livelihood while keeping "outsiders" from intervening. The fort is a physical artefact that reminds deeply rooted and embedded within colonial frontier logics currently still at play.

> Kori: F. David Peat (2005) stated that mathematics should be taught alongside all other subjects, similar to Indigenous ways of knowing, where everything is relative and relational. The Fort, like the school, is premised on barriers and divisions, focusing on differences between what is "inside" versus what is "outside." Subjects learned in school are taught in blocks (both in time and space), disconnected from one another, a process and division also reflected in research and development. But knowings are all interconnected and related. When I set up a tipi I think about the spirit of the trees used for the poles, the angles of the poles, the complex aerodynamics created from the 4-pole structure that

works *with* the strength of the Southern winds, the holistic teachings connected to each of the poles, the woman's teachings of the lodge, the significance of the numbers of pins in connection to the family caretakers of the tipi, while the cosmos represented in the paintings on the upper canvas connects our star ancestors of the meteorites beneath our feet. I am grateful for these teachings and experiences shared with me by Kainai Elder Casey Eagle Speaker and Siksika Knowledge Keeper Treffry Deerfoot, hand to heart, hai hai. Science, technology, engineering, art, mathematics are woven together within the depth of stories and experiential learnings of the tipi. These knowings cannot be disconnected, and when they are, it leads to objective knowledge and therefore appropriation, a tipi is not just a cone (Doolittle, 2006).

Eun-Ji Amy: The divisions and categorisations based on academic disciplines and paradigms continue to prevail. Colonial projects have been possible with the development of science and technology. Through the military and printing press technology, the "New World" was colonized and occupied by settlers. In the STEM curricula, Eurocentric science and math has been inside the Fort while many non-Western based knowledges including diverse Indigenous knowledges have been excluded from the curricula spaces, remaining 'outside' of the curricula Fort. The current STEM education inherently is deeply embedded in Eurocentric thoughts, which continues to promote colonial frontier logics. All of this might not be new for the readers (you). Many Indigenous and ally scholars have been engaged in investigating, thinking about, and critiquing the STEM education as a colonial space.

In his chapter, "A'sugwesugwijig (Meet on Water by Canoe): Learning how to incorporate my Mig'maqidentity in Euro-Western Education", Alex Gray, Mi'gmaq author shares his own academic experience of living in different worlds; in his home community, Listuguj, and Campbelltown and Montreal. He uses a river and water as an analogy of relationship between his own cultural knowledge traditions as well as Eurocentric academic experiences he had "across" from a river. He offers his own way of coming to succeed in both systems by sharing his story of first re-connecting with his community, of which lead to him to create a community for Indigenous students in the field of STEM.

Kori Czuy's "dear big S Science" illustrates knowledge production and validations processes stemming from Enlightenment-based (Eurocentric) science. In this creative prose, Kori highlights the disconnect of Western Modern Science from the land and being human, much like the Fort. She envisions a coming together of Indigenous and Western ways of coming to know science through relationality and community-centered protocol.

Netukulimk is a L'nu (Mi'kmaw) word that describes a process for living sustainably and *tepiaqu* means "enough" in Mi'kma'ki. In their chapter, "STEAM as informed by Netukulimk: Engaging in the radical to consider how to do things differently," Dawn Wiseman, Lisa Lunney Borden, and Simon Sylliboy offer their own stories of teaching and learning to illustrate their ideas in relation to interdisciplinary Land-based teaching, stemming from their reflection on Netukulimk and tepiaqu.

Remembering & Relationality

In using the term "interdisciplinary" approach, we are mindful of diverse stances and definitions attached to the term. We acknowledge the current reality of formal schooling and education that follows different disciplinary approaches in teaching Science, Technology, Engineering, Arts, and Mathematics. Teacher education programs are divided into different disciplinary specializations. K-12 curricula are divided into these different subjects. Interdisciplinary approach allows educators and learners to look at the relations between the already existing division of knowledges and ideas due to the colonial frontier logics. We see interdisciplinary approaches as a starting point to move towards more holistic ways of teaching and learning. A holistic approach that considers different relations of concepts, ideas, one another, and building and sustaining relationships with each other and the Land becomes foundational.

As we write this chapter, we are hearing about the recoveries of bodies at various Indian residential school sites across Turtle Island. Focusing on *remembering and relationality*, it only felt right to reflect, dwell, and sit with this discomfort ...

breathe

Kori: I, along with many that went through the mainstream Canadian education system, learned numbers as abstract and disconnected from the everyday, from body, from story, from spirit. Numbers and their representation as digits and symbols, then categorized as prime, odd, natural, real, whole, irrational ... disconnects them from their relationship to land and especially to the humans that *use* them to explain, rationalize, and answer. To quantify is to control and therefore exhibit power *over* (Bishop, 1988), which often involves a process that removes the subjective and claims objectivity (Anderson & Wagner, 2017).

At Indian residential *schools*, children were disconnected from their names and were replaced with a number. Names are an important part of Indigenous cultures; they represent rites of passage, gifts from the ancestors, accomplishments, and as we all change over the years, so do many of our names. To be called a number, without relation to your spirit, your ancestors or family, your land, without any connections to accomplishments, gifts, or transformations, is an act that removes the human from the person. When I received my Cree name, I was immediately overcome with emotion, I felt more connected with my ancestors from Turtle Island and with myself. Up until then, I was only connected to my patrilineal family name, to Eastern Europe, to the stories of resilience of its journey of integration of it being changed for acceptance, connected to a name that was only half my story. We all have many stories, journeys, and names. A name defines you, and without that, especially if it is removed, you are left as a shell, a mere shadow, leaving a person stripped of their identities (TRC, 2015a, 2015b).

Within the TRC, numbers are used in many ways. They show specificity when exact, but in contrast, statistical numbers can often be vague, rounded numbers, with graphs to compare and contrast, all the while losing specificity (Anderson & Wagner, 2017). The most common number is 150 000, the estimated number of First Nations, Métis, Inuit children that attended residential schools (TRC, 2015a, 2015b). Of the estimated 150 000 children, within the named and unnamed registries, a total of 3125 deaths were recorded between 1867–2000, more specifically, in Saskatchewan, according to the TRC (2015b), there were 566 deaths recorded in the same time–period. The recent numbers being released about the bodies recovered at Indian Residential *schools* are exact and not estimated, and at just one school in Cowessess First Nation in Saskatchewan there are 751 unmarked graves; that number rose to 3125 ... then silence.

215: Tk'emlúps te Secwépemc FN, Kamloops Indian Residential School
104: Sioux Valley Dakota FN, Brandon, Manitoba

161: Zhahti Kue, Fort Providence, NWT

160+: Penelakut FN, Penelakut Island, BC

35+: Regina Residential School, Regina, Saskatchewan

751: Cowessess FN, Saskatchewan

39: Dunbow Industrial School, Alberta

74: Battleford Indian Industrial School, Saskatchewan

35: Muskowekwan FN, Lestock, Saskatchewan

182: Ktunaxa Nation, Cranbrook BC

80: St. Boniface Industrial School, Winnipeg, Manitoba

Each of these numbers 1 2 3 4 ... 180. ... 622 ... . 751. .... 1836 ..... 4126. .... *about* 10,000, represents a human, a child. Each of these children, each number, is connected to more numbers—parents, grandparents, aunties and uncles, cousins, that have been significantly affected by the numbers, the graves, the loss of a loved one(s), and are still grieving. Imagine sending your child to school and *never* seeing them again or know what happened to them.

take a moment to think about what that means

As these numbers inevitably grow, let's try again to keep them exact, give respect to the spirit and the generations of families that each single number represents. Anderson & Wagner (2017) said that using exact numbers shows respect for each individual. There was a ceremony at Fort Calgary (that coincidence isn't lost on me) for the first National Truth and Reconciliation Day, September 30, 2021. Front and centre was a giant orange flag with a list of names, names of those recovered, names said out loud to call them back home, names that were never forgotten by families, names that finally replaced the number they were given at residential school, and names that replaced the number when they were found. That flag was only 2/3 full of names, the rest was empty space that was far from empty, it symbolized the names that are still to be found, reconnected, remembered.

As I (Kori) review this introduction in Winter 2021, and again 6 months later, I search again to find updated numbers on recovered bodies … I search again … the shock of the 215 that shocked the world in May 2021 has dissipated. Now, what is left is merely an aftershock, another ripple in time, recovery after recovery, the ripples are no longer felt, a blindness to the stories arises, another Indigenous story pushed under the proverbial rug. The last number I read was *about* 10,000, ABOUT 10,000, can we take a moment to think about how large that number is? Also concerning is that specific numbers turned into rounded numbers, media got bored even as the search continues. Yet each number represents a healing journey for the previously lost one's family, the beginning of answers, and a reckoning for all. This number will inevitably rise …

Breathe

The sacredness of children is at the centre of Indigenous communities, where every act, action, ceremony, prayer, is centered around the children of the seven generations. So, it is only fitting that it is the children being *recovered* from the Indian Residential *Schools* that have finally prompted the calls to *act*ion #71 through #76[2] (TRC, 2015a), can no longer be ignored, no longer be poorly funded, no longer be questioned. These children are the sprouting seeds, finally being seen, heard, and listened to.

Eun-Ji Amy: I feel uncomfortable. I feel uncomfortable because sometimes there are too many theories and stances around reconciliation. The terms like reconciliation and allies are thrown into the realms of academia and policy world. As I am writing a response to Kori's heart-pouring writing on recoveries of the bodies from the residential school sites, I am finding myself feeling uncomfortable. I am sitting here on my desk in my office in Australia. As a non-Indigenous person, now in Australia, how much does my voice and thoughts count or matter in the conversation around the reconciliation and the Indigenous-settler relations on Turtle Island? How can I move beyond the settler's guilt and do the *real* work, that actually matters?
I am reflecting on the calls to action put forward by Truth and Reconciliation and Council (TRC). As a legacy document and at policy level, these calls to actions may have played an important role in setting the stage or creating a space for Indigenous communities to speak. To what extent did it really help? That is not my place to reflect here. But, I am turning to lots of so-called 'ally's work,' citing the TRC calls to action to be part of the 'reconciliation' movement. For me, the TRC is more than policy, more than calls to action. I have a friend who was helping collect survivor's stories. Soon after he had to take a medical leave due to PTSD. I worked in communities where the very artefacts and legacies of day schools and residential schools were remembered by generations after generations.

As a first-generation immigrant to Canada, I have been benefiting from settler's privilege from different treaties. Because of my settler's privilege and working experiences with Indigenous peoples, now I gained a status in Australia. Thinking about my experiences, existence, and privileges, the TRC calls to action are just something that we 'act' upon. Calls to actions stem from the actual embodiment—bodies, stories, and spirits of inter-generational experiences of resilience of Indigenous peoples across the Turtle Island. It is to be (re)membered. In using the term (re)membering, I reflect with Sandra Styre (2017)'s notion of (re)membered—that these stories and spirits of original inhabitants to be (re)membered in contemporary understanding.

Looking at the etymology of 'member', I realized that the term 'member' is linked with being a part of a body. Calls to actions are not just separate entities. They are part of bodies, the bodies buried on the residential school sites, the bodies of survivors, the bodies of people who have gathered in the circles to share their stories and hearts. As such, in engaging with academic work, citing calls to action, I hope the actual bodies are being (re)cognized and (re)membered. In (re)membering these bodies, I continue to grapple with the questions including how can my academic work go beyond performance allyship? Who gets benefit through this work and what kind of benefits?

With the rise of *cancel culture* we currently witness, I come back to the questions above. There are lots of 'pretentious' intellectual workers within the academia and policy world. I've been advised by many of my Indigenous mentors and there are lots of 'fake knowl-edge keepers' out there. As a non-Indigenous person working in the cultural interface (Nakata, 2007), it is not my place to judge one's cultural identity. However, I remain crit-ical in terms of whom I should engage in doing the *real* work in curriculum work. The *real* work involves (re)membering. Connecting curricula, policy, and pedagogies back to the actual embodied experiences of the stories and experiences of the original inhabitants. The *real* work involves honouring the true protocol from the communities I am engaging with and amplifying the voices from the past and present coming from the embodied experience with the place. Such real work only happens based on the relationships built upon trust. The reasons and privileges that are given to me to work on this book project is based on the trust that was shared with my friends and colleagues. When it comes to the terms like Land-Based STE(A)M or beyond colonial frontier logics, the first thing I should do is (re)membering actual bodies and stories from this place, with the people who actually have intergenerational embodied experiences with Turtle Island. I do not have the capacity to do such real work alone. If I do, it wouldn't be real (re)membering.

In the chapter, "**Kaa kishkaytaynaan taanishi lii Michif aen pimatishichik (We'll Learn About Métis Culture)**," Joel Grant, a member of the Métis Nation of Alberta (Region 3, Treaty 7 territory), reflects on his journey of (re)membering and reconnecting with his Métis heritage. Through his stories, he emphasizes the importance of sense of belonging. He honours various mentors and friendships where he points to moments of reconnecting and the notion of reciprocity.

Amanda Fritzlan writes about her stories from teaching mathematics for urban middle school students in the territories of the Skwxwú7mesh and Tslei-Waututh Nations (North Vancouver, British Columbia) In her chapter, "A

Place-Conscious Approach to Teaching Mathematics for Spatial Justice: An Inquiry with/in Urban Parks," Amanda points to pedagogical implications of teaching mathematics for spatial justice, focusing on Indigenous and colonial histories of public spaces such as urban parks.

In their chapter, **Relationship-based science education: Understanding the Mother Earth through the engagement of head, heart and hands through artful-scientific Inquiry,** Eun-Ji Amy, Hannah, and Ro'nikonhkátste share their own teaching and learning experience through relationship-based artful-scientific inquiry, occurred in B.Ed elementary teaching science course. Artful-scientific inquiry includes research inquiry stemming from different disciplines whilst honouring the language and lived experiences on the land/place. Focusing on educating the head (mind), hearts, and hands, the authors draw upon their own processes of getting to know nature connecting the stories and languages from Kahnawake.

Land–Based

We want to acknowledge the notion of Land to involve the bodies and spirits, beyond *terra nullius*. Land involves actual remains of the ancestors (Kim, 2020), which is different from space and place. Differentiating space, place, and Land is central for us. Such differentiation only is possible through relationship-based conceptualisations of Land.

The title of this book encompasses "Land-based", Interdisciplinary, and STE(A)M accentuates our collective vision of making the conventional STEM education driven by understandings of diverse relationships at play in making sense of the world around us. We have been conversing and sharing our own experiences of dealing with our own colonial frontier logics. We have come to the conclusion that this may be a personal journey but also a collective journey as a community of (un)learning; one can't really move beyond colonial frontier logics alone. We then started to brainstorm the ways in which diverse Indigenous and settler peoples can come together as a collective to continue to grapple with the questions of, and to act upon, moving curricula space towards learning the authentic stories of the Land through diverse ways of coming to know.

Going back to the CSSE conference. Our panel discussion at the conference was held at the University of British Columbia. For some, it may be just a space, without particular memories of events or ties to the space. Space is empty and abstract. However, for people who participated in our session, the venue became a place. A place that created a memory, a place to share ideas. Place involves a process of relationship-building. Events happen in a place. Sharing stories happen in a place. Place holds meaning and memories. For the xʷməθkʷəy̓əm (Musqueam) First Nation, the conference venue is Land. Land involves actual soil, the earth,

Mother Earth—the remains of the ancestors. It requires one to have a kinship-based understanding of the place. So, geographically it may be the same landscape, but as diverse people occupy that same landscape it, but landscape may hold different meanings and relations to each individual, based on their relations to their ancestors, and the memory of the landscape. Therefore, when we utilize the term "Land-based" here, we acknowledge that, for one to consider the landscape as the "Land" to them, one must have the kin-ship based understanding of the landscape—the landscape which actually involves the remains of the original inhabitants' bodies. We can't use the term "Land" lightly. The majority of the contributors of this collective work currently reside on Turtle Island, also known as North America. Specific landscape of Turtle Island for some ("Indigenous peoples") is Land and for some (settlers and new-comers) is may be space or place. To illustrate this idea, we turn to our own positioning in relation to the Turtle Island.

Kori: I was born off the banks of the Peace River, in Northern Alberta, ancestral lands of the Cree, Dene, and Métis, but my family moved us to Treaty 7, ancestral lands of the Blackfoot, along with Tsuutina, Îyârhe Nakoda, and Métis, when I was about 5 years old. But I clearly remember the winters up North, the crisp cold air, the loud cracking ice moving along the powerful Peace River, and building multi-roomed snow forts. Most of all, I remember the ancestors in the sky, the Northern Lights, and being woken up in (seemingly) the middle of the night, snowsuit over pajamas, to lay in the snow in awe at the show in the sky. My mother was born in Big Prairie (now Peavine Métis Settlement) in Great Lakes Country in Northern Alberta, and learned how to sew, of which she is an expert (this is true, despite what she says!), while living with her Auntie Lena Jobin. The Jobin family has long Métis history connecting to Red River and the Battle of Batoche. She grew up with Cree influences of fishing, medicine making (cottonwood), and the occasional Cree word or phrase, all taught from her Auntie and Mother. My father's family were declared "displaced persons" from Eastern Europe when arriving in Canada, altering their last name so to better integrate into Canadian English society. Upon returning to Turtle Island after more than a decade of travelling/working around the world to "find myself," the significance and relationality to this Land only occurred to me when I returned. But during those travels I visited several concentration camps in Poland, Land that screamed in pain, of which I felt even deeper as I read the names of those who died there, familiar names. "Land" has been commodified, treated as an object, but those moments when we reconnect with the Land of our ancestors, or Land where significant moments occurred, that spirit and relationality is felt, the connections between blood memories of body and Land (re)member each other. I have felt these connections, but they are scattered amongst Polish tours or quick trips up North (embarrassingly, its going on a decade since I have felt that blood connection to Land from Turtle Island). I am a visitor to Treaty 7, it is the Blackfoot, Tsuutina, Îyârhe Nakoda, and Métis that have the deeper ancestral connections to these Lands, for me, it is space and place (as Eun-Ji Amy defines so beautifully).

Eun-Ji Amy: I was born in Seoul, South Korea and moved to Winnipeg when I was 16 years old. Winnipeg (Treaty 1 Territory) is the traditional lands of the Anishinabe

(Ojibway), Ininew (Cree), Oji-Cree, Dene, and Dakota, and is the Birthplace of the Métis Nation and the Heart of the Métis Nation Homeland. Winnipeg is for me, will never be "the Land" as I do not have any kin-ship relations to any of these Nations. It will always be a place for me, a place where I formed friendships with the people and all other living relations. I have visited, moved, and lived in different landscapes across Turtle Island. It does not matter how long I stay in a place, any landscape across Turtle Island will be wither space or place for me. I am now in transition to move to Meanjin (Brisbane) Australia, homeland for the Jagara and Turrbal Nations. In this transient lifestyle, where I am continuously moving around, I still find myself grounded. As I know, I truly belong to where my grandmother is buried, Yeon-dong in South Korea, this is Land for me. Having such grounding in relation to my own grandmother helps me to be a better friend to the people and other relations of Turtle Island.

In thinking with the notion of "Land" in relation to the ancestors' bodies, we—as the editors and the facilitators of this collective conversation—want to engage our academic work with our critical minds as well as our heart. As such, we focus on the power of stories and storytelling. Therefore, when we first invited each author to contribute to this book project, we focused on gathering stories.

Myrle Ballard is Anishinabe scholar from the St.Martin First Nations. In her chapter, "**Anishinaabe Kwek Piimachiiwin: Indigenous Women's Anishinaabe Knowledge Systems,**" she explains the connection between the Land (Nature), language, and social relations stemming from the stories from the Elders from the community. Myrle draws special attention to the role of women in transmitting knowledge and traditions.

In Roger Boshier's chapter, "**Home, hoe, horse and hammer?: How to learn from and live on the land,**" he explores the notion of "land-based" learning in informal, nonformal, and formal settings. Drawing from examples from diverse places including Canada, UK, and New Zealand, he shares historical and pedagogical implications of each setting to Indigenous-Settler relation and Indigenous resurgence.

The Trent University TRACKS (TRent Aboriginal Cultural Knowledge and Sciences) program team, Kelly King, Madison Laurin, and Kristin Muskratt, share their learning and teaching experience. In their chapter, "**Fostering Growth Through Indigenous and Land-based STEM Education,**" the authors summarize TRACKS's goal, history, collaborative learning process, and diverse outreach programs for Indigenous youth.

A in STE(A)M

Storytelling and arts-based approach provide opportunities to engage with a holistic perspective of the realities of co-existence and shared lived experience on the Land. The term Arts here encompasses all arts-based approaches. However, we focus on the role of Arts in sharing one's stories. It was through the school practices that creates "universal" narrative of knowledge construction as well as people's experiences that led to marginalisation of Indigenous knowledges within the education space we have witnessed throughout decades.

The Eurocentric model of academia continues to divide and categorize the knowledge and peoples' lived experience into different academic disciplines. In using the term "interdisciplinary STE(A)M," we acknowledge the current realities of the existence of boundaries in categorizing knowledge into different disciplines, science, technology, engineering, arts, and mathematics. We use (A) with parentheses. While acknowledging the reality, our point of conversation is to focus on the role of Arts. Especially the role Arts play in creating an active space for stories to be shared. It is through stories that educational researchers and teachers start thinking beyond the deeply embedded colonial frontier logics in the STEM fields. Stories can bring fragmented information and diverse lived experiences on the Land from the past and present.

> Kori: I was recently asked, is *mathematics a science or an art?* Why do we preface this question as an *either/or* instead of *and*? Why is mathematics mandatory and art or Native Studies an option? The mathematics I learned in school was very much abstracted from my everyday life, and therefore I never did well, nor was I interested in the subject.
>
> It was only when I was doing my Master's degree did I ask the question "what is mathematics?" Of which I have only now begun to realize that like science, art, chemistry, medicine, engineering etc … are all defined similarly, mathematics, and all ways of knowings, no matter what category they are segregated within, is based within *relationships*. Mathematics is the relationships between one number, angle, shape, idea, with another, as well as in relation to ourselves, our understanding and our experiences. 2 + 2 = 4, is about the relationships between 2 and 2, with 4, a set of four then can connect to a relationship to four of something in nature, a four-leaf clover, the four winds, a four petalled flower, the early stages of a fractal within a branch of a tree, the relationships to four have allowed them to thrive and survive. Similarly, art is a representation and interpretation of a sense, an idea, an emotion, a thought, a question … .all expressed in relation to the artist, sometimes with the intentions for others to come into that relationship through interpretation or experience.
>
> Math is relationships.
>
> Art is relationships.
>
> Spirit is relationships.
>
> They are all the same, so why are they so segregated and disconnected, trying to be understood on their own, disconnecting the human relationship to them, the land and spiritual relationship to them?

Eun-Ji Amy: This is similar to the scientific knowledge production and transmission practices. When we teach Nature of Science (NOS) to our students, we tell our students that scientific knowledge is subject to change with the new evidence, that there is no single step-by-step scientific method. However, when we teach students for their science fair, we teach the linear 6–7 steps of the scientific method–where we lose sight of the stories that are being told. We teach the scientific law and theory as the 'truth'–the truth that they will have to be able to regurgitate in order to pass or get good grades in standardized testing. As a science educator, I see the role of (A) in bringing the stories back to learning. I see the role of (A) in putting the learners themselves in relation to all evidence, methods, stories, observations into their own inquiry to getting to know Nature better. In that way, the STEM is no longer a place just for learning specific methods, knowledge, and formulas about Nature. Nature becomes a contextualized place or Land (depending on their own relationships with the place, whether friendship or kind–ship) where students started to see themselves in relation to other beings–including bodies and spirits already existing from thousands of years ago.

Look again at the cover art for this book. As each chapter begins to connect to one another so does each story. The spirit of each story is like a bead. Each bead, retaining its integrity, spirit, and personality, comes together with the other beads, to create a collective piece of art. The resulting cover picture depicts moments from each of the stories, all connected through physical pages, coded PDFs, and connected through the thread that weaves the beads, pictures, and stories together into one.

In the chapter, "**De/colonizing pedagogy and Pedagogue: Science education through participatory and Reflexive Videography**", Marc Higgins shares his experience in Iqaluit, Nunavut. With a "twin–lens methodology", the author engaged in participant-directed videography as pedagogy with Inuit Youth at the Nunavut Arctic College. In reflecting his experience in Iqaluit, he further engages in the topics of Whiteness and Eurocentrism operating in science education.

In her chapter, "**Art-the-garden: Decolonial Teachings beyond disciplinary Frontières**", Julie Vaudrin-Charette employs A/r/tography, explore her own experiences and ideas about STEAM education through art-making, researching, and teaching in relation to learning with the stories told about four plants. Julie shares her learning experiences with plants in the Makakoose Medicine Garden, located in the college she works at, connecting stories from the Elders from diverse regions (Barrier Lake, Kitigan Zibi, Abitibiwinni).

Drawing from their reflection on a co-created workshop involving slow and Indigenous pedagogies, Lorrie Miller and Shannon Leddy write about pedagogical implications of "infusion of Indigenous learning across curricula areas" in their chapter "**An Axiology for making- Weaving slow and Indigenous pedagogies- First Peoples' Principles.**"

Radical Hope & Reciprocity

We have been conversing and sharing our own experiences of dealing with our own colonial frontier logics. We have come to the conclusion that this may not only be a personal journey but also a collective journey as a community of (un)learning; one can't really move beyond colonial frontier logics alone.

Kori: But there is hope, there *has to* be hope, radical hope. As I hear the quiet chirps of the chickadee as I write this at my deck, Canada is *finally* hearing the quiet voices of the spirit of those children. But it is a time to not just hear, but to listen. Those children were seeds, buried to be forgotten, but instead sprouted, now are *seen*. The buried seeds are sprouting, like sporadic Nehiyaw words recently sprouting from my mothers' childhood memories. I am grateful for these seeds, for their resilience and strength.

I recognize my privilege in my lighter skin, having my father's Polish last name, and growing up through Western worldviews and traditions. I recognize that it is my mixed heritage has inspired this work of weaving together multiple worldviews and experiences of math and science. Not growing up with my Métis heritage ancestry has inspired me to learn about myself and my ancestors, to create and maintain relationships with those who have ancestral connections to this Land, to (re)connect with capital L Land and the spirit of Mother Earth, but most importantly taking a step back to support, highlight, make space for, and lift up stories and depth of scientific knowledges within Indigenous communities, histories, and ways of knowing. I am witnessing the stories being told, the complex ancestral knowledge being listened to, the Indigenous science being heard. Relationships are being mended and (re)created. This to me is radical hope.

Eun-Ji Amy: For me, I am not too sure about hope. In Korean language, the word 희망 (hee-mang) may be the closest to hope. 희망 involves imagining the future. One can imagine and predict the future-to-come, in a more positive manner–that is hee-mang. That's the difference between Wish and Hope. Wish may not involve active imagination of the future. There is no action-planned. Perhaps, this is exactly the reason why I am still not comfortable when I hear terms like 'reconciliation', 'decolonization', or 'Land-based' education. Understanding the decades of resilience and labour of the many Indigenous Peoples that went into formulating conversations around these topics, the bodies that are buried across the residential school sites, hearing stories of my friends fiercely joining their healing journeys from intergenerational trauma, I can't use these terms lightly just solely for academic purposes. Thinking about 'Land-based education' in relation to terms like decolonization or reconciliation, I am not even sure if I can participate in imagining the future—doing the 'hope' work. Where would my place be, as a first-generation immigrant to Turtle Island, now moving to Australia, in imagining the future of such Land-based Interdisciplinary STE(A)M education in these places?

For me, hope becomes wish–actionless thinking without understanding the principle of reciprocity. I have been so privileged. I have been given so much wisdom and lessons from many of my mentors, teachers, students, and friends from diverse Indigenous Nations. These learning from my mentors and friends allowed me to finish my PhD and do consultation work in communities–which led me to become an academic in a public institution in Australia. It is my responsibility to continue walking alongside (quietly most of the time, but sometimes loud–when my friends need me to) with my friends–sharing my intellectual, financial, and emotional capitals where they are needed in contributing to creating platforms for imagining the future to happen.

In the sharing place, I continue to listen and learn.

And in this sharing place, through our mistakes and ongoing learnings, we invite you, the readers, follow along on this journey, to listen and learn, with open minds and hearts.

Hand to heart.

Hai hai.

NOTES

1 The Alberta Curriculum re-write, during the publication of this book, has not finalized due to a flurry of controversy.

2 We are not going to explain these calls to action here. If these are not known to you, the reader, it is now your responsibility to do the work, and look them up. We are getting tired ...

REFERENCES

Adams, J., Das, A & Kim, E. A. (2020). The Crit-Trans Heuristic for Transforming STEM Education: Youth and Educators as Participants in the World In S. Steinberg & B. Down (Eds.). The SAGE Handbook of Critical Pedagogies. (pp. 1497–1507). Sage Publications Ltd. London: UK

Anderson, A., & Wagner, D. (2017). Numbers for Truth and Reconciliation: Mathematical Choices in Ethically Rich Texts. *Journal of Mathematics and Culture*, 11(3), 18–35.

Association of Canadian Deans of Education. (2010). Accord on indigenous education. Retrieved from: https://www.oise.utoronto.ca/oise/UserFiles/File/FoE_document_ACDE_Accord_Indigenous_Education_01-12-10-1.pdf

Donald, D. (2011). Fort, colonial frontier logics, and Aboriginal-Canadian Relations. In Ali. A. Abdi (Ed.), *Decolonizing philosophies of education* (pp. 91–111). Rotterdam: Sense Publishers.

Donald, D. (2009). Forts, curriculum, and indigenous Métissage: Imagining decolonization of Aboriginal-Canadian relations in educational contexts. *First Nations Perspectives*, 2(1), 1–24.

Doolittle, E. (2006). *Mathematics as medicine.* Plenary lecture delivered at U. Calgary, 2006. In Liljedahl (Ed.), *2006 Proceedings of the Canadian Mathematics Education Study Group.* Burnaby B.C.

Haig-Brown, C. (2012). Decolonizing diaspora: Whose traditional land are we on? In A.A. Ali (Ed.), Decolonizing Philosophies of Education (pp, 73–90). Brill: Leiden, Netherlands.

Kim, E. A. (2020). Positioning myself in Turtle Island: The storied journeying of a first-generation Korean immigrant-settler to Canada. In E. Lyle (Ed.), Identity Landscapes: Contemplating Place and the Construction of Self. (pp. 152–161) Brill|Sense Publication.

Nakata, M. (2007). The cultural interface. *The Australian Journal of Indigenous Education*, 36, 7–14.

Peat, F. D. (2005). *Blackfoot physics: A Journey into the Native American Universe*. Boston, New York: Weiser Books.

Truth and Reconciliation of Canada (NCTR) (2015a). Truth and reconciliation: calls to action. Retrieved from: http://nctr.ca/assets/reports/Calls_to_Action_English2.pdf Accessed on: 24 October 2020.

Truth and Reconciliation Commission of Canada (NCTR) (2015b). What we have learned: principals of truth and reconciliation. Retrieved from: http://nctr.ca/assets/reports/Final%20Reports/Principles_English_Web.pdf Accessed on: 24 October 2020.

Fort

two sides of a river

river as sustenance
science   story   pride
   tobacco   relationship
Land   ceremony
community   locatedness   interconnectedness

                                                  land as business
                        Capitalization   Capitalism   Science
                             reduced   segregated   curated
                                    taken for granted
                                    count, group, sort

          land/Land   biotic/abiotic

enough

disrupt   academic results

address the tensions

two-eyed seeing

struggles　　　　　　　　humility　　humanity　　disruption

radical hope

mycorrhizal　　together　　dignity

Land as family　　　　land-guaging

living together　　　　reliant on each other

ethics of balance

A'sugwesugwijig (Meet on Water by Canoe): Learning How to Incorporate My Mig'maq Identity in Euro-Western Education

ALEX GRAY

PEMIJAJIGA'SIT: WALK ALONG THE SHORE

Nin teluisi Aleq Wyouche (My name is Alex Gray). From a very young age, I was captivated by science. Whether it was running home after school to watch an episode of The Magic School Bus, or watching tadpoles hatch in the school playground, my interest in science could always be found in some shape or form. What made science so alluring to me, was how it could be used as a tool to observe and admire phenomena around us. My mother, Claudia Gray raised me on the stories of *Glusqapi* and his many extraordinary feats, which instilled in me the notion that our world is full of wonders.

Although I now live in Montreal, I grew up in my home community of Listuguj Mi'gmaq First Nation. Listuguj is in the 7th District of Mi'gmaq Territory that we call *Gespeg* (Last Land), or what is now referred to as the Gaspésie of Quebec. Flowing parallel to Listuguj lies the Restigouche River. Our identity as Mi'gmaq people is tied to this river, and there is as much Canadian as there is Indigenous history connected to it. In 1760, the "Battle of Restigouche" between the English and the French took place on the waters of Listuguj. If you are to walk on the beach of Listuguj today, you can find pieces of porcelain and glass that originate from the vessels that were sunk.

A more recent event involving the Listuguj are the "Salmon Raids" of 1981, which we refer to as *Migwite'tm'nej* (We Remember). These were an attempt by the Quebec government to forcefully stop our people from practicing our fishing rights, and the brutality that the Sûreté Quebec committed in Listuguj is still felt today. Albeit this trauma that was inflicted upon my community, by asserting our rights we can continue to head out to the waters to fish salmon to this day.

While speaking with my mother on the topic of this chapter, she told me a story of how her grandmother would interact with the waters within our territory. When my grandfather was a child, he would often go swimming with his siblings in a brook that we call *Sipug*. His mother would accompany them and offer tobacco to the waters. Curious, my father asked her why she would do that. She replied that she was asking our ancestors in the waters to watch over my grandfather and his siblings as they swam. Our waters are not only a source of sustenance for our people, but are living beings that are part of who we are as *n'nu* (the people/Mi'gmaq).

Across the river from Listuguj, sits the province of New Brunswick and a small rural town called Campbellton. Listuguj and Campbellton are connected by the J. C. Van Horne Bridge, which makes daily travel between both locations possible. Most of the larger department stores, restaurants, and health services exist in Campbellton. So, it is quite common for Listuguj community members to do much of their shopping and business there.

Growing up in Listuguj, I attended our community's elementary school from nursery through to Grade 8. The *Alaqsitew Gitpu* (Soaring Eagle) School is the only elementary and middle school in the area to be specifically for Mi'gmaq students. Here, I was immersed in a system that approached Euro-Western education from an Indigenous standpoint. While we had the typical science, math, and humanities classes, we also had additional classes devoted Mi'gmaq culture and language.

My mother has led the program for the Mi'gmaq culture classes since the school was built in the late 1990s. Her focus in the classroom has been on teaching her students how to create traditional and contemporary Mi'gmaq art pieces while tying in our traditional stories. Each grade level attends my mother's classroom on a weekly basis, and for many it is regarded as an opportunity to escape from the stress of expectations in Euro-Western education. Being my mother's son meant that Mi'gmaq culture classes extended into our home. My parents would often bring me with them to collect materials and medicines for my mother's work. We would pick sweetgrass from beaches in the late summer, collect willow branches in the fall, and a number of other materials from the land throughout the year for preparation for the classroom. Although I was reluctant to participate in these "activities" as a child, probably preferring to be indoors or playing sports, I would later find appreciation for these opportunities of land-based learning.

In Corntassel & Hardbarger's *Land-based Pedagogies and Community Resurgence*, they describe the importance of "[fostering] 'land-centred literacies' that connect us to our past, ground us in our present realities and prepare us for the future generations that will face new and dynamic challenges," (2019, p. 113). My mother was transferring our traditional knowledge of the land and our culture to me so that I could see [them] as the sources of strength and resilience of Indigenous nationhood later on in life.

EGWIJA'LATL: DIP INTO THE WATER

Alaqsitew Gitpu School didn't have a secondary-level education program, so we needed to enroll into a school outside of the community once completing Grade 8. Almost everyone in Listuguj speaks English as their first or second language. So, a large majority of the youth attend English high school in New Brunswick. I also went to an English high school outside of my community. For four years, I was picked up by the school bus every weekday and driven twenty minutes to Campbellton, New Brunswick. I have some very fond memories of my time in high school. But I can't say that it was an easy transition from community school, which had close connections with what was happening at my own home, to a new environment that did not reflect my own culture or home community.

There were some academic issues in addition to trying to adapt to a new learning environment in a city. Issues related to the transitional period persisted for almost two grade levels–until I was able to identify what I needed to be successful in the classroom. During the 4 years I attended high school, I was a student athlete, I participated in extracurricular activities and clubs, and graduated with honours. I was also able to take the right courses to ensure that I would be able to enroll into a postsecondary science program.

Participating the *Eagle Spirit High Performance Camp*, hosted by McGill University, played an important role in developing my passion and goal to pursue a degree in the field of science. The camp for Indigenous secondary students was coordinated by the First Peoples House at McGill. Initially, the camp caught my attention because it was catered towards Indigenous student athletes who wanted to excel in their sport. A few of us from Listuguj caught wind of this camp and decided to apply together. Many of us were accepted, and so we all made our way over to Montreal to participate in the *Eagle Spirit High Performance Camp*.

We completed fitness tests, learned about proper nutrition, and went to meet coaches and some of the elite university-level athletes. While this was a large component of the camp, there was also a large emphasis placed upon exploring the life and programs at the University. We participated in various hands-on workshops led by different McGill academic programs. From those, the most impactful program was led by the Faculty of Medicine who invited us to a facility called the *Simulation Centre*. This centre is where many of the health professional students at McGill develop and hone their skill set to work on the front lines of health care. We were given a glimpse of the program and the lifestyles of these students. We also participated in training sessions at the centre with the guidance of medical students and professionals. This included such activities as learning how to interview medical patients and how to draw blood samples. Of these activities, there was one that truly spoke to me; a workshop on patient care, led by a Mohawk family doctor from Kahnawake. Working with a high fidelity medical mannequin controlled by someone in the next room, Dr. Horn interacted with the mannequin as she would with a hospital patient. She addressed any of the patient's questions or needs and showcased the process of resolving a medical emergency. After the workshop, Dr. Horn presented to us the hardships she faced during her educational journey. She shared with us some of the struggles she went through in order to complete her education program. Her stories demonstrated her resilience and drive in achieving her goal of becoming a physician.

Pursuing a career in medicine had been a dream for myself, and Dr. Horn became my role model. Her story helped me to transition the dream of becoming a doctor into a belief. Overall, the *Eagle Spirit High Performance Camp* left such an impression on me that I returned to participate as a camp counsellor. Working as a camp counsellor, along with other Indigenous students from McGill granted me the opportunity to listen and interact with Indigenous students' experiences in post-secondary settings. Hearing the stories of Indigenous students offered a glimpse into what higher education looks like from Indigenous standpoints. I wanted to do my part in ensuring that future campers would have a similarly positive experience at post-secondary settings.

GETAPA'SIT: SUBMERGE UNDERWATER

After graduating from high school, I decided to continue my studies at McGill University and pursue a Bachelor of Science. At this point, I had been to McGill for three times to participate in the *Eagle Spirit Camp*. So, the campus had become a familiar space for me. I also had knowledge of the resources available to Indigenous students at McGill, which made the concept of leaving home a lot easier.

As is the case for many Indigenous learners, going to the university meant that I would be away from my family and community for an extended period of time. At first, I was excited to be living in the city of Montreal and to begin my studies at McGill. I attended my first few classes, met new people, and began the first chapter of my early adult life. While this all appeared great on the surface, the reality had taken a much darker tone. The high expectations at McGill demanded a strict work ethic to succeed, and although I knew that I had the capability to do so, I wasn't succeeding. Meanwhile, I was also in an emotionally abusive relationship, which took a lot of my time and energy away from focusing on my courses. I began receiving failing grades for the first time in my academic career. Failing grades instilled a sense of dread within me. When my parents asked me, "how is everything going?" I skipped telling them the details of my life and hid the hardship that I was going through. I didn't want them to worry about me. After the second semester, my grades were so poor that I was not allowed to continue my studies. I was being kicked out of the university. The sudden shock of this news was crippling to say the least. In my mind, I not only let myself down, but also my community. I was receiving funding to attend university from my community. I felt that this financial support was wasted on me. I felt an overall sense of guilt and shame and I fell into a deep depression.

While I felt that this was my own burden to bear, I still had a loving family who supported me throughout all of this. After some serious talks, my family understood how all of these events affected my physical and mental health and told me that I should consider moving back home to Listuguj for the time being. I was reluctant at first, but I eventually agreed. Being home allowed me to distance myself from the traumatic experience that was going on in my life in Montreal. However, I still didn't have any idea nor plan on what to do with my life. This is when my mother asked me to come to her classroom in the community school and help her during the week to keep busy. Whenever I visited Listuguj during the school breaks from McGill, I would always make an appearance in her classroom to say hello. It was a place that I considered an extension of my home, so I agreed.

I went to the school every day. I helped my mother organize her materials before her classes and sat with her students as she guided them through their work in the classroom. This carried on for a few months. And eventually, midterms were around the corner. Some of the students expressed their difficulties in science and math. My mother looked at me and asked if I would take a few minutes to help a few students with math and science. From this tutoring experience, I started asking students to bring any schoolwork they were having a hard time with. The tutoring carried on for a few weeks, but I was quickly realizing that a few minutes was not enough. I began staying after school to help students with their homework. After school tutoring became a formal initiative at school. I was spending a considerable amount of time helping students and volunteering around the school. I was subsequently asked if I was interested in becoming a substitute teacher. I immediately jumped on the opportunity with joy. Spending my time in the school carried on throughout the winter. Whether I was aiding my mother in her classroom or substitute teaching, I was in the school almost every day working with students. For me, this was an opportunity to give back to my community in a meaningful way. I was doing what I could to help students get ahead in their studies, but I was still unsure of what I was going to do regarding my own education. At the time, working at the school was a welcomed distraction for me and helped to reduce my overall anxiety.

One day, I was heading to my mother's classroom after the bell to head home. She mentioned that some of the students that I taught earlier that day were commenting on my teaching. Right away, I thought that this was either a negative comment or that they were poking fun at my teaching style. To my surprise, students were telling my mother how passionate I was when I was teaching them. For example, how I wouldn't move on from a topic until I made sure that the entire class understood. They saw my face light up when I spoke about science from a perspective that was beyond just exams or the classroom. That science could be found in simply appreciating how animals adapt to the winter months or how our ancestors created canoes that have remained structurally unchanged to present times. Listening to those students' comments made me aware of something in myself that I hadn't seen for quite some time: they saw my passion for science and learning. It was the piece that I had been missing for a long time. I had become so fixated on my anxiety about my academic performance, that learning became a game of trying to stay afloat. Not only had I lost my passion to learn, but I also lacked a sense of community, belonging, and my own identity in Montreal. My experience in Montreal, attending a colonial learning institution was an alienating experience. I think many other young Indigenous learners may resonate with my story. For example, we rarely see our peoples and culture represented in the campus space. When our peoples are mentioned in the classroom, it is often accompanied by negative statistics that paint a very dire picture of who we are. All things considered; it can be so easy for a young Indigenous student to be discouraged by all of this.

When I reflect on my academic experiences, I think of the stark difference in attending school on and off reserve. In my community, our culture is woven into every experience and found in every classroom. This was almost never a part of the content or let alone the conversation in high school off the reserve. I went from learning and practicing my culture on a near daily basis in school, to having only two courses offered in Mi'gmaq language throughout the 4 years of high school. My interest in pursuing a science degree did not allow me to register for those courses as they conflicted with requirements for post-secondary. Even when there was an attempt to incorporate Indigenous peoples into the subject matter, it was never more than a mention of our ancestors or a grossly inaccurate representation of how they fit into the "Canadian" narrative. To be clear, the high school I attended served a population of students that were a majority non-Indigenous. I can even confirm that there were some fellow students from my reserve who had a very favourable experience in attending the high school. On my most recent visit home, I had a discussion with a childhood friend who shared how he completed his high school diploma by taking specialized and unique courses all centered around technology. Hearing this perspective confirmed to me that there were teachers in my high school who went above what was expected of them, and even advocated for Indigenous students to be accredited for courses that did not follow the educational norms. However, the structure of Euro-Western science education in the school followed a very rigid structure and did not allow for different ways of learning. This made for a difficult transition for students coming from Listuguj, who are used to learning science with cultural content embedded within. While I would eventually find my stride, many of my Indigenous peers did not. I would often see that those who were repeating courses or terms were usually the same Indigenous students that rode the bus with me to high school each morning. This system succeeded in making me feel that my identity was no longer the focus, and I consequently whitewashed myself to better fit into the academic model that was being presented to me. In other words, I compartmentalized my culture separately from my concept of academic achievement because it was being demonstrated to me that they were mutually exclusive.

What I am now beginning to understand is that this adopted assumption of separating my Indigeneity from Euro-Western learning is just one example of the complex issues that colonial frontier logics has presented to Indigenous peoples (Donald, 2009). When I reflect upon my own experience with colonial frontier logics, the river and very landscape of my home tells a story. My community was on one side of the river, holding my sense of identity and culture. This was the place where I was encouraged to see things from a Mi'gmaq perspective. Across the water was where the majority of my academic learning took place; where my understanding and thinking was shaped to fit the Eurocentric academic model. Between these two locations, I had spent a large majority of my young life in one place or the other. This also extended to my mental and emotional identity, never before thinking that I could bridge the two. The saving grace of all of this, was that each day I would return home and subsequently to who I am as a *Mi'gmaw* (Mig'maq person).

By the time I left home to pursue post-secondary studies, I no longer had that place of grounding, my community. I remember in one of my first-year courses at McGill, fellow students were discussing their academic backgrounds and the topic of graduating grade averages came up. I managed to graduate from my high school with honours, but when I mentioned my average, they gave me puzzled looks. I was then asked how I managed to get into Sciences as the cut-off was apparently higher than my grade point average. This would be the first time of many that I felt as if I didn't belong in the university. I was dealing with imposter syndrome in the classroom. Not having my home accessible to me anymore meant that I was questioning my own identity, and I quickly became lost in my life at that moment. Perhaps this was subconscious, but what I came to realize was that I was lacking a sense of community and belonging in the University and to the city of Montreal. I knew that I could reach out to the Indigenous student resource centre at the university. But, at that time, I couldn't pinpoint how important and crucial a sense of belonging to a community was to my overall well-being.

NATAWA'Q: KNOW HOW TO SWIM

Eventually, I began to take the steps to return to my studies. I had an understanding of the herculean task I had ahead of me, however, I now had rekindled my passion for learning and rediscovered my identity. I took the steps to reapplying to university, and within a few months, I received word that I got back in. I would spend the next few weeks preparing for the move, but I also took the time to appreciate my home and family as I would again be leaving it all behind. However, it was not goodbye. As encapsulated by the Mi'gmaq word, *nemultes*, which closely translates to "I will see you again", my home was always going to be where it stands. I just had to remember to take it with me in mind and spirit.

The return to university was not a sound transition, in fact, there were still times where I would either just get by in my courses or even had to repeat some of them. Because this would be the first time that I actively engaged in my courses, I was becoming increasingly aware of the challenges that are posed to many Indigenous students across academia. In the time that I returned as an undergrad, I learned that there was only one additional Indigenous student enrolled in my program. This lack of representation across the student body only worsened the initial feelings of impostor syndrome I had experienced. I felt my perspective and learning was not welcomed at this institution. As intimidating as it was being one of the few Indigenous students attending an institution that may as well have been colonialism incarnate, I was reminded that I had access to support systems that did not reflect the Eurocentric cultures and values. There were some initiatives and spaces, including *First Peoples House*, that were reserved for Indigenous students that also had Indigenous representation at the level of leadership.

After being introduced to *First Peoples House* through the *Eagle Spirit High Performance Camp* all those years ago, I finally returned. It was there where I began to slowly meet and get to know the other Indigenous students at McGill. After learning of peoples' names and Nations, I quickly began to learn that my hardships and perspectives were also shared with these other students. Alienation, the pressure and stress of having to represent Indigenous perspectives in non-safe spaces, these were all things that we were going through in our individual programs and studies. For the first time in my academic career, I realized that this overbearing sense of dread and unbelonging was not unique to myself.

In the institution, we all felt unease being ourselves and representing our peoples and identities; my hardships were validated. In my mind, I had been carrying my shame and guilt for my past attempt in university like a steel beam across my back. And it was at the *First People House*, where I suddenly felt the weight lifted, and I could stand taller.

In this space, we as the Indigenous students supported one another through the daily challenges of attending McGill. I had finally found my community, and it brought me a sense of home and purpose that I was lacking prior. The other piece that helped me truly call *First Peoples House* my home was when we were given the opportunity to participate in ceremony and cultural practices. *First Peoples House* would host cultural events like storytelling with Elders, feasts with traditional foods, and even an annual campus wide Pow-Wow. Through these opportunities, I suddenly felt empowered to share the things that I had learned from my family. For the first time since leaving my home, I was granted space to be able to share my identity as a Mi'gmaw. All of those times spent in the woods with my mother and father picking medicines, being taught Mi'gmaq words, and hearing my mother sing our songs had finally come to the forefront of who I was; my identity was whole again.

Although my time as an undergrad at McGill did not come without its hardship, I was able to learn how to navigate the space while staying true to who I am. I slowly started to have the confidence in myself to engage in the classroom. I maintained an active presence at the *First Peoples House*, remained involved with some Indigenous initiatives that would later open doors for me, and in January 2018 I completed my B.Sc in Physiology. For all the challenges I had faced during this time, I finally was approaching the end of this chapter. A moment that I still hold very dear to my heart, was when my mother had gifted me an eagle feather after the graduation ceremony. Of all the people who supported me through my degree, my mother was always the one who witnessed my efforts. She comforted me in my defeat when I was kicked out of university. She provided me with tools when I was on the path of finding myself. Not once had this woman given up hope on me, and when she gifted me that feather, she told me that I had earned it for all of the sacrifice and growth I had gone through to be where I was on that day. For those of you who understand the degree of honour it is to receive an eagle feather, you can imagine the pride that I had felt in that very moment. Although I did not think of my actions as deserving of this, my mother reminded me of the great triumph I attained that day.

ELISGNUET: WEAVE/BRAID

A few months after completing my degree, I was asked if I would be interested in working for a program that was soon to be launched at the University, the *Indigenous Health Professions Program* (IHPP). What was enticing to me about this initiative was that it was formed from years of consultation with Indigenous community members and other workers from the health services. Indigenous people were at the head of leadership for IHPP. One of these leaders said they had known me since my undergraduate studies and knew of my commitment to the *Eagle Spirit High Performance Camp*. The facilitation of the camp was being handed over to the IHPP and was to be rebranded as the *Eagle Spirit Science Futures Camp (ESSF)*. Little to say that all of this combined made it a very alluring opportunity for me, and I accepted an offer to work as the IHPP Outreach Administrator.

Considered as a rebranding and starting new, my objectives with the ESSF camp were very open in the beginning. I started to teach students science from the focus of health because I felt that it was easier for me to apply my knowledge into my teaching practice. While this was an engaging and fun way of approaching the work, I quickly felt as if something was missing. Here I was, as an Indigenous person, working with Indigenous youth from a very Euro-Western perspective. I felt as if all of my training and studies at University did grant me a toolset, but it did not sit well with me to be approaching science from the very same lens that negatively affected me as a secondary student.

Something that I would draw inspiration from for my science teaching, came in the form of our first annual ESSF Camp. While ESSF was to maintain its focus on promoting post-secondary studies for Indigenous youth, our program transitioned it towards concentrating on the health science programs offered at McGill. In addition, our community partners had expressed their concern that students are not interested in science as a topic in the classroom. The very objective of the IHPP is to get more Indigenous health professionals in the field, but the reality is that students who wish to pursue such health professions must perform well in science. In addition to the usual content of the camp, we were implementing a science curriculum that would be approached from Indigenous ways of thinking. We worked with an Indigenous science teacher from a nearby community (Kahnawake), and from this the curriculum was formed with land-based teachings and storytelling woven into the experience. I was in awe at what could be accomplished when science was approached in this way. The students were engaged and the inner child in me finally felt that his identity was being reflected in the very lessons that were being delivered. This was the perspective of science I had been searching for my whole life. For the first time ever, it was clear to me that the knowledge I had inherited from my mother and our lands as well as the lessons I had received in the classroom could weave together harmoniously.

Now after having a few years of applying the two-eyed seeing approach, I have been able to find comfort in the idea of teaching scientific knowledge from Indigenous ways of knowing. A highlight of such teachings was a moose organ dissection I led in my community's school. The scientific content knowledge was rooted in mammalian physiology and anatomy, but I wanted the activity to be carried through in a Mi'gmaq fashion. The moose organs were donated by community hunters from their hunt in the fall. They were also present during the dissection and were able to talk to the students about their process of hunting and how it provides traditional food to the Elders and community at large. The day prior to the dissection, I taught the students about human anatomy using the structures that they would be working with the very next day. My reasoning for this was to demonstrate how we are connected to our four-legged cousins through the commonalities of our bodies. We all have a four-chambered heart, and two lungs to breathe and circulate life into our bodies. Tobacco was given to honour the moose for his sacrifice, and the organs were given back to the land when we were finished.

All in all, my educational journey has been a path of self-discovery, learning and unlearning, and overall empowerment as an Indigenous educator. There have been times when I felt utterly lost in my own identity, just as there have been times when I feel my pride as a Mi'gmaw swelling within my being. This story that I tell of my life with science continues to unfold as time goes on. However, it is now clear to me the role of science within this story. Too often, we look towards Indigenous ways of knowing as a supplement to the static process of Euro-Western learning. As Indigenous peoples, we come from cultures that are based in the oral tradition and knowledge of our lands. It is because of this, that I do not say that our Indigenous ways "compliment" or "enrich" the scientific method. In fact, I say that the Euro-Western scientific method is a tool to help tell the stories that we as Indigenous peoples want to convey.

REFERENCES

Corntassel, J., & Hardbarger, T. (2019). Educate to perpetuate: Land-based pedagogies and community resurgence. *International Review of Education, 65*, 87–116. https://doi.org/10.1007/s11 159-018-9759-1

Donald, D. (2009). Forts, curriculum, and indigenous Metissage: Imagining decolonization of Aboriginal-Canadian Relations in Educational Contexts. *First Nations Perspective, 2*(1), 1–24.

dear big S Science

KORI CZUY[1]

AUTHOR'S NOTE OF RE-PUBLICATION:

The last few years have been difficult, finishing my PhD (during a global pandemic), beginning a new job (during a global pandemic), hearing of recoveries of unmarked graves at Indian residential *schools*, controversies of identity politics, and science becoming intertwined with politics … just to name a few. These challenges have brought about difficult conversations, personal growth, and simultaneous reconnection and disconnection, but all in all have prompted a more intensive journey to better understand myself and my ancestors. Throughout the timeline of writing, publication, and republication, my journey to better understand my ancestors has changed. I realized my mixed identity as defined by my mother and by the government didn't coincide. The piece of plastic issued by Métis Nation of Alberta connects me with my Jobin ancestors, aligns with names like *halfbreed* that mother remembers being called growing up, but aligns less with her recently recalled memories of the Cree language, her childhood stories of living on/from the Land, or her requests to acknowledge our ancestors. We are all on our own journey, and mine is ongoing and complicated.

I wrote this piece in 2017 and it was only published in 2021! Even despite the pandemic, the barriers and obstacles continuously enforced within strict publication formatting and its limited parameters for creativity is shocking. Needless to say, my poetry for now lives outside of a Word Doc. In this re-publication, footnotes have been retained instead of publication suggested endnotes, because I believe it allows for a closer connection between those references and stories.

****b r e a t h e o u t****

my b r e a t h .. o u t .... spreads my s p i r i t ....

b r e a t h e i n

i breathe in ... your spirit

connections are made

dear big s Science

little s science here[2]

remember me?

we shared a tea back when you were *scientia* remember?

scientia latin for knowings

knowings not knowledge

the -ing connects generations past with generations yet to be

alive with spirit

we had more in common when you were scientia

may i (re)fresh your memory?

(re)member ...

little s science

i am knowings relationships connections interconnections spirit connections

i am equity and equality not hierarchy or dominance

as little s science i embrace all within the circle all equal all with

openness to give and receive

open heart & open mind (Eagle Speaker)

little s s c i e n c e

each letter equal

no dominance over hierarchy over control over

little s s c i e n c e

working together towards a whole knowings ~~about~~ with

the whole

all

all my relations

respecting our responsibility to our universal family (King, 2003)

generations past present future

knowings from our teachers with 2 legs and 4   with scales   feathers   fur
claws   hoofs

knowings from all that move and breath in and out

and all that don't

coyote   kokum   fern rock turtle naa––mooo[3] father star

knowings

knowings from all with all   together

big s Science
you have changed
you now act like a Proper Noun
your big S hovers over little i
you exert your big s Science when you need to
like you have control of your   Capitalization
        Capitalism
big s Science you have changed   from   *scientia* knowingssss
say it   feel it   *scientia knowings*   it dances with my tongue
knowledge big K   K   K   the sharp sound cuts
carves out and discards the little   the subjective
the stories under the rocks and with the kokums
Knowledge   big   K   reduced   boiled   down   cut
apart   curated   generalized
knowledge of everything
needing to know everything
assuming everything can be   has to be   *needs* to be known
knowledge – "something to be possessed and accumulated" (Peat, 2005, p. 55)
a static noun
it feels like block   hard   ridged   cold   white
teachers fill students with these blocks of knowledge
students receive knowledge fill the bucket (Freire, 2005/1970)
        only to regurgitate it
to fill in square after square of limited choices on a test

where do the blocks go after?
maybe some are (re)collected
used to reinforce the rigid standardized checklist for post-secondary admission
(re)collected   copied   pasted   inserted onto the mounting wall that separates the
academy
from community

from humanity

from knowings s s s

big s Science you have a crush on   a lust for   over   knowledge
a crush on   a lust for   over   the power of Knowledge   big K
knowledge to control with its power over nature
hierarchy of man over nature (Hatcher et al., 2009)
control her
make her obey
put her in a box if she doesn't

the box
a metaphor for big s Science
with walls to separate ||| the observer ||| the observed |||
subject ||| object
but the box of Science knowledge is closed   separated   segregated
to be observed   looked at
watched
hard Science
   hard to argue with   negotiate with
   hard to crack
   hard to have a cup of cedar tea with and get to know it better
   hard to woo
   hard to connect with through those walls
big s Science
you say the human distorts
alters knowledge with its impurity   faults   imperfections
      big s Science   you make me feel dirty   unwanted
                              i am a disruption to your academic results
                        remove the human from the equation   the tests

            get into   the box

      hold on .... who's in the box   what's in   the box?

little s

big S

the walls are too high   distorting

who/what is in   who/what is out?

breathe

you watch

i feel

i feel your eyes   your gaze

all you seem to have is your eyes

they have become your superpower

enhanced through the microscope and telescope (McLuhan, 1962)

you mechanize   technosize your vision

reduce   enhance   focus

generalize

finalize

you lost the human in your eyes

big s Science, your hierarchy controls anything and everything lower case
how did you also manage to control my senses?
plato   descartes   galileo
vision allied with reason logic   the ideal
sight became distanced from other senses (Classen, 1997)

increasing with the rise of the modern age

photos lens instruments observation – the surveillance of Science

>   memories   stories   experiences   spirit
the gaze of the scientist = acquisition of knowledge (Foucault, 1973)

big s Science   you removed my ears   fingers   tongue
am i just a specimen to you?   put the sample in the box
those samples/senses are for humans
lower to the ground
less evolved
without the shiny plasticity of modern white
damnit descartes you have given me cartesian anxiety (Bernstein, 1983)

darwin   land bridges   primate evolution   lucy
the creation stories of big s Science
value   support   (re)produce power and dominance through hierarchy and
privilege
struggles are *seen* as weak
the strong are rewarded by passing on their genetics
the winners   thumbs up
skywoman   turtle island   connection between cosmos and earth mother
collaboration and respect for all within the circle
the creation stories of little s science
with   not over
together
no losers
within the stories with the land   struggles are vital
inniee matueeee- a hearty grass of the prairies
to reach the sun   had to struggle through the
packed down earth from the buffalo
to grow   thrive   and provide food for the buffalo (Crowshoe, 2019)
the circle is complete
the circle is interrupted
the great buffalo hunt   sparked by big c Colonization
to remove many native people's vital source of   food   shelter   clothing
ceremonies …
to control
to win over the land and animals and people of the land
without the struggle
without the buffalo
the inniee matueee disappeared (Crowshoe, 2019)

with big s Science's vision as its dominant sense

how is there so much blind trust in this capital S?

discovery   conquest   exploitation   distribution   appropriation (Tuhiwai
Smith, 2012)
you have colonized   commodified   my connection with earth mother
"a forceful d i s r u p t i o n
of a set way of knowing and doing" (Eagle Speaker, 2019)
uprooted the people whose lives   spirits   generations live with amongst
the land

amongst little s science
"remove the indian"
"disconnect them from the land   civilize them"
"educate them with big s Science"
uprooted   roots exposed   vulnerable
do you know how it feels when your roots are snipped?
when your nerves are snipped   exposed raw?
    transplanted
    replanted
    unstable unsteady without the security of the foundation of your roots
    no choices   must depend on   rely on others
dominant others with mechanical roots
must rely on others for support (Donald, 2009)
                              the lucky ones still have some roots

        sometimes only as memories   blood memories

        maybe not your own memories   from generations past
                              that travel through spirit
                not flesh
                              through story

            not the written word
                              through radical hope
(re)planted into the garden of big s Science
man-made garden
without spirit
but with fences   walls   barriers to control and observe
                              my spirit retreats
transported to new home

                    unsafe

                              marginalized
analyzed

numbered

     named (re)named

you call me fnmi
it hurts my lips and my tongue to say
        say it out loud … FNMI
twisting my name into poky letters
i don't recognise myself anymore
fnmi
who am i? where am i? when am i?
you might as well give me my number back at least then i am individual
not generalized into a group without meaning

     can we work together? again?
  (re)concile   (re)ignite (re)member
    back when scientia little s science and big s Science
  were friends?

        were relations?

     (re)concile this truth

      truths

not big t Truth
truths plural
  isn't that what this is all about?
  your truth my truth our truths the truths of the land
    can we (re)member (re)validate truths lost through
colonization capitalism commodification   capitalization
  the truths that have been removed stolen grabbed kidnapped
  taken from their home land mothers
  uprooted
  (re)rooted
  never the same
      but earth mother has nurtured these severed roots
        kept their hidden truths safe
          like mothers do

    to (re)connect to

truths acquired through knowings

protocols

all ways of knowings have protocols to acquire truths

and taught   passed on   by knowledge keepers

big s Science
knowledge keepers— professors   academics
recognized by peers (community) through certifications   publications   pre-
sentations
written words
to guard the protocols   methods   rigor   the written word

little s science
knowledge keepers are
respected   recognized   by people within the community
gifted with relationships   knowings
to keep knowings safe   passing them on through relationships created
through trust
reciprocity
knowledge keepers
certifications aren't on paper   but gifted through song
story   bundles   ceremony
visions

protocol

community accepted processes

for coming to knowing truth   / truths

(Czuy & Eagle Speaker, 2019)

big s Science
your protocol   as accepted by the (academic) community
begins with tuition   commodity
expectations of submissions   papers   written words
submit   obey   submit
with teachings to be received
teaching towards   teaching to
down to

student being filled with knowledge
students being taught   taught down to
knowledge
as protocols   mandated processes to acquire truths
scientific method   standardized method
      to question   reduce   analyze   generalize   replicate   present
      verification   validation?
      recognition only through
      publication of words   numbers   statements

          capitalized   statements

    publication / words on paper/ words on screens   = validation
statements of big k Knowledge
all working towards upwards   increasing   validation
certification
paper   stamp words   titles   those sought after letters
b   a   b   s   c
   m   a   m   s   c
        p   h   d
permission granted to control over
to enhance the rigor
uphold   withhold the standards
… of the box

          little s science
        protocol   as accepted by community
    as a way to acquire truths with and from   knowledge keepers
   protocol towards the journey of being entrusted with knowings
          begins with an offering
          tobacco[4]
      reciprocity respect   gratitude to earth mother
      reciprocity respect gratitude to knowledge keeper
reciprocity   respect   gratitude to generations past   present   future
      smoke from tobacco[5]- connection to the spirit world
   tobacco–   a contract to speak the truth from heart from spirit
      humility of what we do not know
      openness to the unknown
      tobacco
   to nourish life   relationships   knowings   truth

these strong relationships   the roots that anchor the process of protocol
relationships   stories   truths

story   of   the
land   cosmos

of past generations
creation stories
moral stories
everyday stories
miracle stories

stories told   not read
listened to   felt   experienced   connected with
not recorded
no pen   no paper   no tapes   no   computers
senses and memories and experience as personal records
stories are gifts of truth
truth from generations of experience on

with
amongst
within
the land
knowing from stories through generations of knowings
stories dripping with knowings
soaked in understanding of land
heavy with spirit
experienced   learned   only through story
no paper   no written words to see
only spirit   senses   experience
(Archibald, 2008)

not unlike big s Science

stories as knowings   passed down   orally   through generations

mistakes   humility   learning   experiential learning

repetition

inference prediction

observations

pattern   recognition   prediction

truths as
subjective
contextual
sensory
based in relationships
with land
community
self

spirit

but

stories are held close now

severed roots of stories are guarded   cared for

but often retreat

trauma from past

intergenerational trauma

trust is now difficult

trust is now earned

stories are gifted   NOT taken

NOT told TO   but WITH

big s Science
you come into my "lab"   earth mother
you don't even knock   or offer tobacco   or pray
                    where are your offerings
                i didn't hear "all my relations"
            did you say your name for creator to hear?
you just walk in   assuming your (academic) credentials and titles   give
you access
        access to gather truth   but little t truth

you try to understand   little s science truth
through the protocol of big t Truth   big s Science
you impose your recording devices   you don't even ask
        instruments that distance your body   your senses   your cul-
ture
        from   little t truth
    trying to gather   sample   collect   quantify   validate   little
t truth
scientific method
but ... these protocols are for big s Science   big t Truth
why do you assume you can use your protocols
                        to find my   little s science   little t truth
                                        to little s science
this is like if i entered your lab   in my kokums moccasins and ribbon skirt
with ceremonial tobacco to offer as protocol to create a path towards truth
                        to build relationships of reciprocity with the
                        professor   your   knowledge keeper
                                        to gain trust
                                        in the hopes
                                        sometime
                                when the time is right
                                to be gifted stories
                        of truth   of big t Truth
                your truth but through my protocols ....

when i am allowed to thrive in my little s science world

i am never alone

spirits from generation past and future guide me

animals   plants   the cosmos

teach me about survival   tell me stories through their actions   interactions

my senses and observations and intuition build creativity from the land

i observe

i experience

i make mistakes

i laugh

i learn

i make connections with the spirits of the land   animals   trees   cosmos

connections   knowings

i am never alone

we are always together   equal   in the circle
often this is not reality

little s science has been living in a flip floppity world   every sunrise

little s science has been living without its human

without its senses

without its animal ancestors

without its medicines

without its knowledge keepers amongst the cosmos

awkward
lost
confused
uprooted   rerouted

big s Science i hope like me you feel like we are (re)acquainted now

friends   co-creators

together on a journey

(re)membering   creating

looking past   experiencing present   hoping

radically hoping

dreaming towards the next sunrise

together

big s Science

can we learn through both our worlds?　　together?

maybe sometimes i can use my little s science protocols
and teach you little s science truths

eager to learn　　offering tobacco　　creating relationships

maybe visit during the new moon　　my spirit is stronger in the dark

my senses to knowings are stronger in the dark

open to coming to knowing

together

open to new

open minds　　open hearts

all my relations

* * * *breathe in* * * *

AUTHORS NOTE:

as i write this, type this, my spirit continuously fights with the controlling powers of colonialism through technology
through the typed word
assuming, no, pushing its controls over which words are to be "proper nouns"
forced capitalization
controlling which words therefore can exude more power through capitalization
it takes me three or four attempts to type a small i
i　i　i　i what if i don't want that capital letter? it's for the big S scientists, the big P philosophers
it's just me　　my ancestors　　and all my relations
but we are all equal　　we will sit together, equal, amongst the lower-case
　i fight the controls of formatting
　font 12
　calibri　　new times roman　　arial
　meh. they don't amuse me, caress my sight, make me want to read more

every u j p is the same
         copied   pasted
yawn
sometimes, a letter or a word feels different to me, it needs a different font.
i want to listen to its request
spaces   tabs   alignment   pre-formatted, difficult to re-format
automatic bullet points raises my anxiety

        let's step back for a moment …. "bullet" points? really?
i type this to share
    to share with eyes and ears and hopefully hearts and spirits of those who
blindly abide by these parameters   restrictions   rules   boxes
for those who wouldn't think to question
who are scared to question

            who have never questioned …. …. ..

NOTES

1 Grateful acknowledgment to *Springer Nature* for permission to reproduce copyrighted material: Czuy, K. (2021). dear big S science. *Cult Stud of Sci Educ.* 16, 357–372. https://doi.org/10.1007/s11422-020-09983-7. Reprinted by Peter Lang with permission of *Springer Nature*. All rights reserved.
2 this is my journey of (re)connection with my roots, buried deep within the soil, hidden for safe keeping …
3 From Piikani Elder Reg Crowshoe, na-moo, meaning bee in Blackfoot, it's not just a word, not just written, but when said, naaa-moooo mimics the sound of a bee, the interactions of spirit between the bee and the ear, hearing and connecting to the song of the bee (p.c.)
4 Tobacco is a sacred plant to many First Nations and Métis communities across Turtle Island (North America), but not all. Different communities have different traditions based on the land they are on. This reference is from oral teachings from Blackfoot and Plains Cree Elders.
5 The smoke here refers to burning the tobacco, either in a smudge (a ceremony, often including other sacred medicines such as sweetgrass, sage, or cedar, burning in an abalone shell where the smoke takes the prayers to Creator), or in a ceremonial pipe. Smoke is not inhaled.

ALL MY RELATIONS AND INSPIRATIONS

Archibald, J. Q'um Q'um Xiiem. (2008). *Indigenous storywork: Educating the heart, mind, body, and spirit.* Vancouver, BC: UBC Press.
Bernstein, R. (1983). *Beyond objectivism and relativism: Science, hermeneutics, and praxis.* Philadelphia: University of Pennsylvania Press.

Classen, C. (1997). *Foundations for an anthropology of the senses*. Oxford UK: UNESCO Blackwell Publishers.

Czuy, K., & Eagle Speaker, C. (2019). Critical braiding approach to ethno(mathematics). In M. A. Peters (Ed.), *Encyclopedia of educational philosophy and theory*. https://doi.org/10.1007/978-98 1-287-532-7_648-2

Donald, D. (2009). Forts, curriculum, and Indigenous Métissage: Imagining decolonization of Aboriginal-Canadian relations in educational contexts. *First Nations Perspectives*, *2*(2), pp. 1–24.

Eagle Speaker, C. (ongoing). Recognized Elder of the Kainai Nation, Treaty 7. Ongoing oral teachings from 2012 to present. Teachings on Treaty 7, Calgary, Alberta, Canada.

Eagle Speaker, C. Recognized Elder of the Kainai Nation, Treaty 7. (2019). Oral Teachings- October 8, 2019. Teachings on Treaty 7, Calgary, Alberta, Canada.

Foucault, M. (1973). *The birth of the clinic* (trans. A.M. Sheridan Smith). New York: Random House.

Freire, P. (2005/1970). *Pedagogy of the oppressed*. New York: Continuum.

Hatcher, A., Bartlett, C., Marshall, A., & Marshall, M. (2009). Two-eyed seeing in the classroom environment: Concepts, approaches, and challenges. *Canadian Journal of Science, Mathematics and Technology Education*, *9*(3), 141–153.

King, T. (2003). *The truth about stories: A Native narrative*. Toronto, ON: House of Anansi Press.

McLuhan, M. (1962). *The Gutenberg galaxy*. Toronto: University of Toronto Press.

Peat, F.D. (2005). *Blackfoot physics: A journey into the Native American Universe*. Boston, New York: Weiser Books.

Tuhiwai Smith, L. (2012). *Decolonizing methodologies: Research and indigenous peoples* (2nd ed.). New York, NY: Zed Books.

CHAPTER FOUR

STEAM as Informed by Netukulimk: Engaging in the Radical to Consider How to Do Things Differently

DAWN WISEMAN

Bishop's University
Ktinékétolékouac (Sherbrooke, QC)

LISA LUNNEY BORDEN

St. Francis Xavier University
Nalikitquniejk (Antigonish, NS)

SIMON SYLLIBOY

St. Francis Xavier University
Eskissoqnik (Eskasoni, NS)

Netukulimk: The use of the natural bounty provided by the Creator for the self-support and well-being of the individual and the community. Netukulimk is achieving adequate standards of community nutrition and economic well-being without jeopardizing the integrity, diversity, or productivity of our environment.

(Unama'ki Institute of Natural Resources, 2016)

FOREWORD: BY WAY OF PROVIDING SOME CONTEXT

The editors of this volume have challenged us to consider the following questions:

- In what ways could we do *real* work in moving beyond the colonial frontier logics in STEAM education?
- Also, how might the conceptualization /relationship with 'Land' enhance the settler-Indigenous relations in the STEAM curriculum space?

We welcome the questions. They get to the roots of the work we do in teacher education, and alongside teachers and students from First Nations, Inuit, Métis, and Canadian communities in various local places. The questions have made us think more deeply about how to engage in ideas with others. For us, this paper is part of a conversation-in-progress-for-many-years; for the reader, it may represent an early foray into new ways of thinking. We thus invite the reader into this conversation with hopes they will consider what the terms and ideas might mean in the contexts of the relationships of their own locations of teaching and learning. We lay out where we are at in our own knowing, being, and doing with respect to key terms and ideas within the questions, as well as some of the larger ideas that inform our work. We do not claim these ideas are new or "ours", only that they have been generative and fruitful for us. Lisa[1] and Dawn[2] have come to them over a long period of still-ongoing learning in and alongside First Nations, Inuit, and Métis people, Peoples, and communities. For Simon[3] the ideas are more family/iar, some learned, others are a part of who he is and where he comes from as L'nu. The ideas form the base of ongoing collective conversations with others about teaching, learning, land/Land, and STEM/STEAM, where Netukulimk is circulating together with related ideas in Cree, Michif, and Dene while we attempt to share thinking about them in English.

As the plurality of languages indicates, our work requires negotiating the realities of what is contemporary Canada. While we work towards "liv[ing] together in dignity, peace, and prosperity on these lands we now share" (TRC, 2015, p. 13), in order to move toward more mutually sustainable relationships between Indigenous and Canadian people, Peoples, and communities, we start with the truth of Canada as a colonial state founded on genocide (Palmater, 2014). This positioning leads us to understand that while English is generally the language for sharing our work, English words are a colonizing force frequently inadequate for describing ways of knowing, being, and doing that come from the Land in the way that Indigenous languages do. So, we question and play with language as means of trying to come to deeper understandings and to see if the language proposed to us is in fact useful to our work and to the people and places with whom we work (Lunney Borden & Wiseman, 2016; Wiseman & Lunney Borden, 2018).

As we have laid out in more depth elsewhere (Lunney Borden & Wiseman, 2016; Wiseman et al., 2020), we understand both STEM and STEAM are terms more and more frequently being taken up in education discourse as descriptors for the interdisciplinarity that arises in commitments to inquiry and problem solving (often in relation to and emergent from Land or land-based learning). We have tensions with these terms. They often seem caught up in and framed by European Enlightenment traditions, and the ongoing contemporary entanglement of these traditions in capitalism, colonialism, patriarchy, neoliberalism, and White supremacy (Grosfoguel, 2011); instead of ways of coming to understand the world in all of its abundance. To counter, or perhaps bypass these entanglements, we posit that what might be labelled STEM/STEAM in English are actually artefacts of commitments to inquiry and problem solving–in teaching, learning, or just living–in–relation to particular local places and all the land, air, water, and other relations therein. We hope that this positioning of the terms is one way of overcoming colonial frontier logics (Donald, 2009) in STEM/STEAM education, but we are open to discussion and more learning on this point (and all the others we present).

We also distinguish between land and Land. The capitalized version, per Zinga and Styres (2011), is an active teacher, rather than a passive objectifiable place where learning might occur, as is often the case in initiatives labelled "land-based learning". Land instead

> indicates an identifiable geographical location where animate earth, air, and water come together with all the beings (human and other-than-human) who exist and have existed in physical, emotional, intellectual, and spiritual relationships with and within that place/Land. This is *Land* that is "storied" (Donald, 2009, p. 129), *Land* that speaks (Basso, 1996), *Land* that can be listened to (Hogan, 2000), *Land* as first teacher and pedagogy (Zinga & Styres, 2011), and *Land* that has birthed countless generations and accepted them back in death as a part of itself (Wilson, 2001). (emphasis in original, Wiseman, 2018, p. 338)

This understanding of Land, and its difference from land as resource (as land is often conceptualized in Canada), helps us to further understand Canada as a nation-state mapped over and attempting to subsume hundreds of Indigenous nations for whom this Land is and always has been home (Manuel & Derrickson, 2015). And so, we refer to "what is currently Canada"–as the phrase is sometimes taken up in Indigenous futurisms (Keene, 2018)–to suggest Canada both is and was something different than politically portrayed in the contemporary world and can be again.

Finally, while we do locate our work in what might be called STEM or STEAM education, we tend to see the problems and challenges of STEM/STEAM education as a "microcosm of the macrocosm" (Cajete, 2006, p. 249). That is, the problems and challenges in STEM/STEAM reflect those that humans (and all living beings) face on much larger scales in terms of living (Watt-Cloutier, 2021). So, while the following chapter may be read as a commentary on STEM/STEAM education (and it is), it is also a commentary on current dominant global discourses and what has got the global "us" here[4]. As such, the chapter does not present an argument, as much as an offering of ideas for thinking with and alongside for considering how to overcome divides by doing things differently.

INTRODUCTION

This chapter was first conceived in the relative peace of Summer 2019 when we were planning collaborative work with educators in a number of First Nations communities in different parts of what is currently Canada. That project considers what might be labelled STEM/STEAM teaching and learning as conceived in and by First Nations communities, very much emergent from and connected to land (in some cases) and/or Land (in others) and specific local contexts. Goals for the project include challenging taken-for-granted understandings about where and with whom expertise in STEM/STEAM teaching and learning lies and demonstrating how STEM/STEAM might exist apart from conceptions of land as capital resource (Donald, 2019).

Nearly a year and half beyond the initial chapter conception, that research has been reconceptualized due to the ongoing global disaster of the COVID-19 pandemic, as has the final version of this chapter. In fact, the world from the current vantage point seems incredibly different than it did in 2019 (but this is perhaps a failure of our own understandings and imaginations). Just comprehending the events of 2020 has made writing challenging–wildfires around the world (Werner & Lyons, 2020; Witze, 2020), planes being shot from the sky (York et al., 2020), ravaging hordes of locusts (Kennedy, 2020), and pipeline protests that (once again) laid bare the inability of settler peoples and governments in Canada to meet their treaty and human obligations to Indigenous peoples (Wente, 2020). And these events were just the precursors to the kind of pandemic not seen in a century, accompanied by protests of police brutality perpetuated against Black and Indigenous communities (Breen, 2020; Fine, 2020; Martens, 2020), widespread political and civil unrest (Lopez-Martinez, 2020), and insurrection in the United States (Associated Press, 2021; BBC News, 2021). If we believe that education supports young people in understanding the world, we wonder what kind of world we are currently trying to understand, and how the localness and care of the kind of STEAM work we try to undertake might support that understanding.

The editors of this volume have asked us to think about land and STEM/STEAM teaching and learning by engaging critically with Donald's (2009) challenge to move beyond the "colonial project of dividing the world according to racial and cultural categorizations which serve to naturalize assumed divides and thus contribute to their social and institutional perpetuation" (p. 20). While we find the challenge timely, we note that categorization is a foundational part of conventional STEM/STEAM teaching and learning. It is taken up from the early years when young children are taught to count, group, and sort things by colour, size, or number, and asked value questions such as "Why have you grouped things this way?", "Which thing is bigger?", "Which group has more?", "How many more?", and "How might we share this fairly?" In Western school systems as they are currently constituted, these early childhood activities seem natural and innocuous; playful (sometimes) pedagogical precursors to deeper thinking and skills that young people spiral back to over and over again as they move through school. However, these questions and activities have the potential to engrain the idea that everything can be grouped and categorized, and that the world becomes manageable when grouped and categorized, thus opening the door to the deep divisions Donald (2009) seeks to overcome. Here, we raise this point not to focus on division, but to make visible the European colonial underpinnings of Western school systems in order to consider what becomes possible when educators (including ourselves) recognize problematic foundational assumptions to what they do and seek to do things differently.

In a paper completed slightly pre-pandemic (Wiseman et al., 2020), we ended by noting that the kind of struggles evident in the world at that time–particularly genocide, pipelines, natural gas storage, Indigenous sovereignty, and the events of 2020–are the kind of issues we have engaged in with K-12 students and other learners in the teaching and learning of STEAM. These explorations of real world events lead to emergent curricula that is focused not on disciplin(e)ing students but on living together with young people in teaching and learning (Wiseman, 2016), and facing each other as human beings (Donald, 2018), such that students develop a deep sense of community, locatedness, and meaning related to their learning. The explorations require a willingness to engage deeply with the interdisciplinary aspects of STEM/STEAM, and to confront the ongoing realities of colonialism, White supremacy, patriarchy etc. that play out in STEM/STEAM and life. This kind of teaching and learning also requires difficult conversations about how to build relationships with each other and the entities that sustain us as human beings (Donald, 2013). This chapter begins from that ending (Wiseman et al., 2020) to suggest we are collectively in a moment of learning where it is possible to consider who we are in relation to the lands where we teach, whether we relate to these places as land or Land, and how they might sustain us in teaching, learning, and living. We do not pretend to have hard and fast answers to these questions, but we think the questions are worth considering in terms of what they might compel us to do (King, 2019), particularly in a moment where we might have the opportunity to pause and deliberately choose to do things differently.

Framing the Work in the Radicalness of Netukulimk and Tepiaq

As we began to think of these questions collectively Lisa was reminded of Netukulimk, an L'nu (Mi'kmaw) word used in contemporary circumstances as similar to the English term "sustainability" but meaning quite a bit more. Netukulimk describes a process for living sustainably in Mi'kma'ki (Prosper et al., 2011), a territory covering parts of what are currently Nova Scotia, New Brunswick, Prince Edward Island, Québec, and Newfoundland. The concept emerges from and is interconnected with the Land (Zinga & Styres, 2011) of Mi'kma'ki. (We recognize that in other places similar concepts in other languages may exist that reflect the kinships and interconnectedness in those Lands.) For L'nuk, Netukulimk is related to the idea of tepiaq (enough) when enough is considered as the basis for maintaining a healthy, thriving world that extends well beyond the human to include a deep sense of reciprocal kinship with other-than-human relatives, including air, water, and land. Both Netukulimk and tepiaq recognize that health and thriving are relational and reciprocal; that any living being is only as well as the elements that support its living, and that living beings have a responsibility to maintain the wellness of those elements (MacMillan & Prosper, 2016). Netukulimk and tepiaq thus provide a possible counter narrative to prevailing European colonial and capitalist frameworks within what is currently Canada, a state built on the spoils of resource extraction (Wente, 2020) that seems hell-bent on extracting those resources to the point of extinction (both of the resources and of all the relations that are disrupted by that extraction). It is within this framework that we consider whether Netukulimk and tepiaq provide a way to think differently about STEM/STEAM in teaching and learning. Moreover, given the current push to "reopen the economy" (Valiante, 2020) under ongoing pandemic conditions in the pursuit of more and more capital (Klein, 2007) at the expense of living beings, we wonder if enough might be a radical idea, and whether engaging with the idea of enough in teaching and learning renders educators radical.

We take up radical (Barnhardt, 1988) in the botanical and etymological sense of "going to the root … and fundamental to existence" (p. 880) that nonetheless may also be interpreted as in favour of extreme change. We find support for this position in the work of Potawatomi botanist Robin Wall Kimmerer (2013, 2021), who takes up the idea of enough by asserting that that Earth provides everything we need and asking how we root ourselves in kinship with the world through a "radical kind of gratitude" (2021) that sustains this state of being. So, how does this play out in our thinking?

Well, the multiplicity of the root metaphor of radical allows us to re/consider ideas at play in our work. It allows us to acknowledge all traditions have different roots–cultural and philosophical heritages to the ways in which different people and Peoples conceptualize knowing, being, and doing in the world–and to wonder, alongside Donald (2009), how we can both acknowledge the importance of those different roots and simultaneously move beyond what divides us. Here we find some inspiration in another botanical term, *mycorrhiza*–the interrelationship between plant roots and fungi where "neither the fungus nor the plant can flourish without the activity of the other" (Tsing, 2015, p. 138)–to consider how difference tends to focus on either/or options, whereas moving beyond divides asks us to consider the plurality of both/and (Wiseman, 2016). The mycorrhizal both/and can be applied to the interdisciplinarity of STEM/STEAM teaching and learning; to pedagogies that focus on what it means to live well and live well together; to how land/Land might be related; etc. It allows us to think a little differently and perhaps follow up that thinking by doing things differently, by taking up ideas of enough. Doing things differently can be seen as extreme, or radical, but if the difference is rooted–as Netukulimk and tepiaq are in Mi'kma'ki–perhaps radical becomes *mycorrhizal*, and can be understood in ways that allow Canadian and Indigenous people and peoples to "live together in dignity, peace, and prosperity on these lands we now share" (TRC, 2015, p. 13).

We realize it may seem like we are just playing with words here, but such play is sometimes necessary in overcoming divides, and our interest as educators is in overcoming them. Over the last 30 years, a good deal of research, much of it undertaken in the Canadian context, has demonstrated that science, technology, engineering, and mathematics education (both as standalone disciplinary areas in teaching and learning and as more integrated STEM approaches to teaching and learning) are challenging locations for overcoming epistemological, ontological, and cosmological tensions between Indigenous and non-Indigenous conceptions of the world and the way it works (Aikenhead, 1997; Bartlett, Marshall, & Marshall, 2007; Brayboy & Castagno, 2008; Cajete, 1994; Kawagley & Barhardt,1999; Little Bear, 2000; MacIvor, 1995; Michell, 2005; Sutherland & Henning, 2009). Even where there is good intention and good will in attempting to address the tensions, the process of overcoming the divides is challenging.

Given such intransigence, we find some hope in the radical. And we posit that enough, and the related idea of Netukumlimk, might be radical enough ideas for STEM/STEAM educators to learn from; ideas potentially at the root of fundamental change within STEM/STEAM teaching and learning that may allow for overcoming divides. We believe that Netukulimk and tepiaq can provide a framework for STEM/STEAM that is rooted in a way of being in the world that is profoundly anti-capitalist and anti-colonial, and thus fundamental to existence.

In taking up such positioning in our work, the question we often get from teachers (both Indigenous and settler) is "What does it look like?" The "it" is often not clearly defined but refers to some sense of teaching and learning of STEM/STEAM in ways that honour and draw from Indigenous ways of knowing, being, and doing in attempts to get at the both/and. In fact, "What does it look like?" is the research question we posed in relation to the project upon which this chapter was supposed to be based. But because that research has had to be reconceptualized, in the absence of the newer stories we thought we might share, the authors offer some older ones that led us to dig further into the question in the first place.

Netukulimk in STEAM: Attending to Different Stories

"What does it look like?" This question is not easy to answer, because ideally "it" refers to teaching and learning connected to and emergent from land and/or Land. Land changes. It is different at different times of the year. It is different in different places. So, the answer is related to both space and time, who is taking up the work, and how they engage with/attend to land and/or Land. Here we mean land as both family/iar (Wiseman, 2016) local spaces in community, and Land as described by Zinga and Styres (2011) that speaks in the languages it taught to First Peoples, what Glanfield, Thom, and Ghostkeeper (2020) refer to as land-guaging. In the first sense, small "l" land is accessible to students and teachers everywhere as a location for STEAM teaching and learning connected to and emergent from local places with which they have relationships. In the second sense, large "L" Land is described by Ghostkeeper (in Surkan, 2018) in terms of mutually sustaining relationality "I am the land, I am the environment, there is no separation between us and the land" (para. 5). In both our teaching and lived experiences, these two conceptions land/Land are not binaries, or even different ends of a spectrum, but ideas that co-exist and may be in conversation with each other in different uptakes of STEM/STEAM teaching and learning, or in uptakes of STEM/STEAM teaching and learning in different contexts. To illustrate our meaning more concretely, we offer the following three stories, one from each chapter author, that engage with Netukulimk and tepiaq (or related ideas) and land/Land in different ways.

STORY ONE–BLUEBERRIES

Our first story is from Simon. It is a personal recounting of introductions to Netukulimk alongside family and the family/iar. Simon originally wrote this story in his Master's thesis (Sylliboy, 2020) and felt it was an appropriate story to share here as we conceptualized this chapter.

The word Netukulimk was introduced to me over the last decade. However, the teachings and way of being have been deeply woven into my roots as L'nu. There are many lived experiences when I think about Netukulimk, but the significant memory that often comes to mind is from when I used to pick blueberries with my grandparents. When I was younger, my grandparents often took my cousins and me to an area called Mountain Road in Eskasoni. This location had a small blueberry patch where the community would go to harvest. I remember getting frustrated with hand picking these berries, and I asked my granddad if we could just get rakes and collect all of the berries, so we do not have to come back. He then questioned me what would happen if we took them all. I answered, "Other people won't be able to eat them", and, "Other animals would be hungry." He then told me that other people and animals need these berries and that it is essential only to take what we need. After having this conversation with my granddad, this was something that I always thought, taking only what we need, and making sure we leave enough for others. I always reflect on that memory of picking blueberries, which reminds me of the importance of living in a balanced and sustainable way.

STORY TWO-OH MY GOODNESS

Our second story is from Dawn's dissertation (Wiseman, 2016). It comes from a conversation with a Canadian science educator from Alberta about the definition of science within provincial curricula, and how that definition structures thinking in ways that may break down relationality as well as the understanding that both the animate and inanimate (as defined in science) support life and living.

Tracy: [The Alberta science programs of study for science] talk about science as being knowledge about the natural world. That's like really the overarching definition, right. It's knowledge about the natural world. Well, what's the natural world? ...

Dawn: Do you think [the definition] implies there's a non-natural world?

Tracy: Oh, absolutely. Yes, there's natural things and then there's not. That we talk about ... biotic and abiotic. Living. Non-living. And to me, I feel like in that definition [of science] there's living and non-living. And so we talk about how the non-living things affect the living things, but we never look at the other way round. Right? We never talk about how the living things are affecting the abiotic pieces. And I think that is part of that definition too. Right? [But if the definition of science is] knowledge about the natural world, we don't care about the things we wouldn't consider natural.

Dawn: But see now here's ... so that, that split, that living/non-living split. It's one of the big ... areas where things can bump up against each other [when the curriculum tries to engage with Indigenous ways of knowing, being, and doing]. Right?

Tracy: Sure.

Dawn: Because in many Indigenous perspectives there isn't, there isn't that split. Right? Um, because things are seen in relationship. And so-and again I throw this out just for the sake of conversation, because we talk about how those non-living things impact the living. What would be gained, um, in the kids' learning, or in our teaching, if we did acknowledge those, you know, that's not a one-way relationship?

Tracy: Oh my goodness. Don't you think that if we had actually addressed that—would we be in the position where we've extracted our natural resources to the point of, like, they're going to run out? We've actually exploited natural resources. I mean they are counted as abiotic in the curriculum. And so, if you actually talked about them as being in a reciprocal relationship and had an appreciation for those things, I don't know if a lot of the environmental issues that we're facing would be as severe, potentially. (p. 236)

STORY 3–IT'S NOT YOUR TURN TO SWIM

Our third story is from Lisa's work with Kelly, a former teacher now principal, at a Mi'kmaq school, and Tabetha, her colleague as they spoke about culturally based inquiry projects. Both women are from the community of Sipe'kne'katik and teach in the community school.

Kelly described how she explained Netukulimk to her students when on a field trip to a local beach. "When you take something you give back. That's what I explained to the students for Netukulimk. That's why when you take anything from the Earth you're going to give back as an offering. Tobacco is giving thanks to the Earth." She described how, while on the field trip students were collecting things on the beach such as shells and small rocks, but "for anything they brought back with them they made a tobacco offering to give back to the Earth." She explained how one student, in considering Netukulimk, said, "Well, I don't really need this so I should put it back in its place."

Kelly and Tabetha also talked about how important traditional Mi'kmaw stories are in helping Mi'kmaw children understand the ways of Netukulimk. Kelly shared how, as children, they were always told "It's not your turn to swim." This is one of many stories passed down through generations of Mi'kmaq as guidance for how to live well in the world. The tradition was that you do not swim until after the lightning strikes the water after the summer solstice. The rationale behind this story is that fish were laying their eggs and swimming would disrupt their habitat and thus disrupt an important food source. Kelly said these stories "get us asking questions. Elders hold a lot of knowledge but you don't really know it unless you ask them. They're not really going to volunteer it unless you ask them." Tabetha also shared that this wisdom even lies in the language and the thirteen moons which in modern times have become the names of the months. She explained, "just even in that Mi'kmaw word [for the months] it describes what is happening in that month. That's why I really wish I had the language. It's so rich. That one word just tells you so many different things."

Thinking About the Stories in Relation to Netukulimk, Tepiaq and the Radical

We could sit with each of the stories above and link the teaching and learning present in them to provincial and territorial curricular and teacher competency requirements and learning outcomes. There is rarely any question for us that in engaging with land and/or Land the educational outcomes expected by various polities in what is currently Canada can be met (Lunney Borden & Wiseman, 2016; Wiseman, Onuzcko, & Glanfield, 2015). In the stories we share, we see thinking that directly connects multiple elements of STEM/STEAM, including understandings of life cycles, food webs, ecosystems, reproduction, weather and climate, stewardship and sustainability. But we also see stories that go considerably beyond curriculum competencies and outcomes to get to the root of what is fundamental to existence, or the radicalness potentially offered to teaching and learning by Netukulimk and tepiaq. In doing so, the stories offer thinking about what might change or how things might be done differently in terms of STEAM teaching and learning.

Simon's blueberries story demonstrates intergenerational teachings in relation with Land that both originate far back in time and live in the contemporary world. While he learned from his grandfather, the story is located in the blueberry patch in Mi'kmIk'i waiting to be told to each generation that visits the patch to gather berries together and bring them home to share with family. The blueberry patch provides a place where an entire philosophical and spiritual approach to the world lives and teaches about self in relation to others—where the idea of others extends well beyond the bounds of the human to include animals and plants (and other elements not necessarily evident in the story). It also teaches about living together in good relations by respecting the limits of the blueberry patch to provide for all the beings that depend on it. The ethics of sustainability and balance, Netukulimk and tepiaq, are never taught directly, but learned in relationship with others and the Land. Having met and heard the words first with family, Simon now brings them into his own teaching related to STEAM (and other subjects) in ways that bring Land into the classroom and the classroom to the Land.

Tracy on the other hand struggles with received wisdom and definitions that are unsatisfying when considered in relation to the realities facing us in the contemporary world. She challenges ideas underlying curriculum-as-plan (Aoki, 2005) to contemplate what happens when land has the potential to become Land and an active agent in teaching and learning. Her struggle identifies inadequacies of the philosophical roots of science curricula, and points to the need for deeper considerations of the breadth of relationships attended to in science (and by extension STEM/STEAM) teaching and learning. She does not have a familial or Land-based location for grounding ideas similar to those Simon learned from his grandfather, nor is she located in Mi'kma'ki. Nonetheless, she recognizes the lack of and need for an expanded ethics of sustainability and balance that take living beings' mutual relationships with each other and land into account. She now brings this into her teaching to make space for a plurality of understandings in STEM/STEAM teaching and learning.

Kelly and Tabetha struggle with curriculum-as-plan in ways similar to Tracy while making a conscious effort to reclaim the lessons of Simon's blueberry patch. They strive to create space for their Mi'kmaw students to have a connection to the land (or possibly Land) that, due to the impacts of colonialism and the legacy of the local residential school, they themselves were not always afforded in their early lives or schooling. For Kelly and her students, the lessons presented themselves not in a blueberry patch but on a beach where shells and stones provided the means for her to teach about Netukulimk and tepiaq, showing how the ideas and language are rooted in all the Land of Mi'kma'ki. In thinking about how these lessons emerged, Kelly and Tabetha connect to stories held in the Land and told through generations about living in good relations. They have come to recognize the pedagogical purpose of these stories, such as the ethics of knowing when it is not time to swim, as something they can continue to bring into their teaching to address STEM teaching and learning while simultaneously challenging underlying colonial values, like categorization, that often structure how STEAM is taught.

We note that these teachers are located in different contexts and are at different points of their learning. Simon, Kelly, and Tabetha are working in Mi'kmaw Kina'matnewey schools where a decolonizing approach to teaching and learning supported by teaching with and alongside Land is welcomed and encouraged. Though the two schools might vary in their enactment of decolonizing pedagogies, these community schools were built with the intention of thinking differently about education and centring a Mi'kmaw perspective (Paul et al., 2019). On the other hand, in Alberta, Tracy is attempting to make space for similar work but in a much more colonial location–the Canadian public school system where the commitment to unlearning colonialism can be spotty (at best) to non-existent. Yet, despite different contexts, there is similarity in their processes of unlearning and yearning to do something different, not solely for their students but to honour the relationships which sustain life. As Simon would point out, this gets to the heart of Mi'kmaw spirituality and sense of what is sacred. The spiritual and sacred are not ideas we often address in STEM/STEAM, but perhaps they are present as described by Tuck and McKenzie (2015) who note that "in Indigenous worldviews, relationship to land are … familiar, and if sacred, sacred because they are familiar" (p. 51). In our thinking (as it currently stands), the key term here is familiar–or perhaps family/iar (Wiseman, 2016). It underlines the type of kinship and reciprocity inherent in kinship at play in Netukulimk and tepiaq, and perhaps missing in conventional STEM/STEAM teaching and learning founded on categorization and the clear delineation of divides.

COMING BACK TO THE QUESTIONS

There should be nothing more familiar to us, or to the students we live with in teaching and learning, than the world. But we wonder how conventional STEM/STEAM teaching and learning—and particularly programs promoted as STEM/STEAM focused on robotics, coding, or pre-packaged problems—with no connections to people, communities, land and/or Land, support that family/iarity. Not that we dismiss how cool coding and robotics can be; they certainly play significant roles in helping us understand places humans cannot easily go and perhaps relations to be explored in the future (NASA, n.d.). There is a place in teaching and learning for multiple approaches to STEM/STEAM (Lunney Borden & Wiseman, 2016), and perhaps this is radical.

For example, what we shared in Kelly's story does not capture a later part of the conversation where she discussed the work she does with robotics in her classroom, yet her decisions of what to prioritize and when is always being negotiated in relation to multiple desires and discourses that define the wholeness of STEM/STEAM education. This also seems to be Tracy's struggle with curricula–she is questioning how it reflects wholeness as opposed to the categorization and division that allow for extractive thinking. Kelly, Tabetha, Simon, and Tracy are thus engaging in the process of negotiating a desire to live well in the world— Netukulimk and tepiaq—while meeting the requirements of curricular demands. In doing so, they find space to push beyond the extractive boundaries of STEM/STEAM to embrace a way of working differently by being with and considering land and/or Land. Here they are radical by rooting STEM/STEAM teaching and learning in values such as Netukulimk and tepiaq as opposed to assuming that STEM/STEAM is value free (or rooted in colonial and associated values).

As we stated at the outset of this chapter, if we believe that education supports young people in understanding the world, we wonder what kind of world we are currently trying to understand, and how STEM/STEAM might support that understanding. The stories told by Simon, Tracy, Kelly, and Tabetha bring us back to the choices that we make when we engage in STEM/STEAM teaching and learning, and the possibility of actively questioning and resisting the notion sold to us by provinces and territories that STEM (if not STEAM) is about economics and survival in an increasingly competitive world (Lunney Borden & Wiseman,

2016). Their stories show us that different decisions are possible when we root ourselves in what the land and/or Land makes available to us, when we connect rather than silo our teaching and learning experiences. Here they are radical by rooting STEM/STEAM teaching and learning in context, in land/Land, and the mutually dependent relationships therein.

In their opening up of STEM/STEAM to multiple stories, Simon, Tracy, Kelly, and Tabetha enact and put into action Donald's (2009) challenge to move beyond the "colonial project of dividing the world according to racial and cultural categorizations", they question "naturalize[d and] assumed divides" and provide means for challenging "their social and institutional perpetuation" (p. 20). In fact, like Donald (Donald, Glanfield, & Sterenberg, 2012) in more recent work, they move from considerations of what divides us to a more *ethical relationality* that opens a space where we might live together in teaching and learning, facing each other as human beings (Donald, 2018). Simon recognized this as aligning with Mi'kmaw spirituality; we see the sacredness in each other and in the land/Land that sustains us. We thus suggest that taken together, the experiences they shared underline the importance of opening up STEAM to multiple stories that might allow a different ethic of care to be the starting point for teaching and learning.

Closing Thoughts

So what? We are roundly terrible at conclusions, mostly because the work is always in process, but it is always possible to provide closing thoughts that may open up new thinking. We recognize something mycorrhizal in our work as we see this chapter building on and from previous writing and projects–there is clearly something at the root of what we are trying to get at. For the moment, the question we want to ask is the question we think Simon, Tracy, Kelly, and Tabetha might be trying to answer in their stories: *What is the point of STEM/STEAM teaching and learning if not to solve problems in the world?* As potential answers to the question, their stories can help us understand our current contexts of pandemic or climate change in a different light. Consider for example, the (by now) familiar mathematical undulations of "The hammer and the dance" which depicts a flattening of the infection curve over multiple waves of virus outbreaks, such that the capacity of any health system is not overwhelmed (Pueyo, 2020). Our understanding is that in order for this approach to work well, the transmission rate of COVID-19 must be below 1; that is each infected person will transmit the virus to less than one other person. In places where the approach appears successful (e.g., Thompson, 2020), it seems that the number of cases has been close to zero for one to three weeks before restrictions on interactions and opening up have been lifted. More locally, most Canadian provinces (not to mention the United States) have not met such benchmarks prior to reopening, and it seems like market forces are at play: that is, there is the neoliberal and capitalist assumption that there will be costs to the pandemic and that we are willing to pay those costs in people's health and lives. But, as the stories we shared show, we could start from a different assumption that attends to a different set of values, such as ethical relationality. In this case, the same curve might exist, but it is based in a whole different set of behaviours that are more in line with valuing people's lives, and—in the case of schools at all levels—focusing our thinking on the radical idea of how to support all our students instead of dividing attention to consideration of different modes of delivery. These stories also lead us to see how countries worldwide are reliant on each other. What happens when the ties that bring us together are broken down, and how our very survival only comes in sharing resources like personal protective equipment, understanding of viruses, and the risk and hope of vaccine development, manufacturing, and distribution.

We have posited that being radical is about "going to the root ... and fundamental to existence" (Barnhardt, 1988, p. 880). We have provided examples to highlight what this might look like; Netukulimk and tepiaq—or more local but related ideas—as a framework for STEM/STEAM that roots teaching and learning in a different value system—namely one that honours the notions of sustainability through recognition of interconnectedness and interdependence of all our relations, including the other-than-human relations. Being radical nonetheless may also be interpreted as in favour of extreme change and we are okay with that. We believe the message of the current times is that the radical might be what is needed and living a value of enough in relation to land and/or Land and context might be a good starting point.

NOTES

1 Lisa is a professor of mathematics education at St. Francis Xavier University in Nalikitquniejk, Mi'kma'ki (Antigonish, NS). She is a descendent of the earliest Acadian settlers to arrive in what we now call Canada and has worked with and alongside Mi'kmaw communities for nearly 30 years. She currently holds the John Jerome Paul Chair for Equity in Mathematics Education. For more detail see Lunney Borden & Wiseman, (2016).

2 Dawn is an Associate Professor in the School of Education at Bishop's University. Bishop's is located on unceded Abenaki Territory in the area known as Ktinékétolékouac (Sherbrooke, QC). She is an immigrant to what is currently Canada who has been working to understand and take up her treaty obligations for over 25 years. For more detail see Lunney Borden & Wiseman (2016).

3 Simon is L'nu (Mi'kmaq) from Eskissoqnik (Eskasoni) in MI'kma'ki, where he taught science at Alison Bernard Memorial High School. He is currently a PhD student at St. Francis Xavier University working on explorations of Land, language, and science.

4   With the recognition that the actions which have brought "us" here are rooted in Western traditions and colonialism that have given rise to capitalism, neoliberalism, and market logics that strip us of relations, kinship, and meaning beyond the economic (Donald, 2019).

REFERENCES

Aikenhead, G. S. (1997). Towards a First Nations cross-cultural science and technology curriculum. *Science Education, 81*(2), 217–238.

Aoki, T. T. (2005). Curriculum implementation as instrumental action and situational praxis. In W. F. Pinar & R. L. Irwin, Curriculum in a new key: The collected works of Ted T. Aoki (pp. 111–123). Mahwah, NJ: Lawrence Erlbaum.

Associated Press. (2021, January 11). US House introduces impeachment article accusing Trump of 'incitement to insurrection'. *CBC News*. https://www.cbc.ca/news/world/us-house-democr ats-session-trump-1.5868379

Barnhardt, R. K. (1988). Radical. In R. K. Barnhardt (Ed.), *Chambers dictionary of etymology*. H. W. Newyork, NY: Wilson Company.

Bartlett, C., Marshall, M., & Marshall, A. (2007). *Integrative science: Enabling concepts within a journey guided by Trees Holding Hands and Two-Eyed Seeing*. Two-Eyed Seeing Knowledge Sharing Series, Manuscript No. 1,. Cape Breton University, Sydney, NS. Institute for Integrative Science & Health. http://www.integrativescience.ca

Basso, K. H. (1996). Wisdom sits in places: Landscape and language among the Western Apache. Albuquerque, NM: University of New Mexico Press.

BBC News. (2021, February 5). Russia expels European diplomats over Navalny protests. https:// www.bbc.com/news/world-europe-55954162

Brayboy, B. M. J., & Castagno, A. E. (2008). How might Native science inform "informal science learning"? *Cultural Studies of Science Education, 3*(3), 731–750.

Breen, K. (2020, May 29). George Floyd death: Trudeau condemns anti-Black racism in Canada as protests erupt in the US. *Global News*. https://globalnews.ca/news/7002419/george-floyd-trud eau-racism/

Cajete, G. A. (1994). *Look to the mountain: An ecology of Indigenous education*. Rio Rancho, NM: Kivaki Press.

Cajete, G. A. (2006). Western science and the loss of natural creativity. In Four Winds (Ed.), *Unlearning the language of conquest: Scholars expose anti-Indianism in America* (pp. 247–259). University of Texas Press.

Donald, D. (2019). Homo Economicus and forgetful curriculum: Remembering other ways to be a human being. In H. Tomlins-Jahnke, S. Styres, S. Lilley, & D. Zinga (Eds.), *Indigenous education: New direction in theory and practice* (pp. 103–125). University of Alberta Press.

Donald, D. (2018, November 19). Unlearning colonialism: Holism and ethical relationality as cultural forms that can heal us. Public lecture, Bishop's University, Sherbrooke, QC.

Donald, D. (2013). Foreword. In A. Kulnieks, D. R. Longboat, & K. Young (Eds.), *Contemporary studies in environmental and Indigenous pedagogies: A curricula of story and place* (pp. vii–viii). Rotterdam: Sense Publishers.

Donald, D. (2009). Forts, curriculum, and Indigenous Métissage: Imagining decolonization of Aboriginal-Canadian relations in educational contexts. *First Nations Perspectives, 2*(1), 1–24.

Donald, D. T., Glanfield, F. A., & Sterenberg, G. (2012). Living ethically within conflicts of colonial authority and relationality. *Journal of the Canadian Association for Curriculum Studies, 10*(1), 53–77. http://jcacs.journals.yorku.ca/index.php/jcacs/article/view/34405

Fine, S. (2020, June 2). Nunavut RCMP officer to face criminal probe after video after video shows Inuk man being hit by truck. *The Globe and Mail*. https://www.theglobeandmail.com/canada/article-rcmp-ask-outside-force-to-investigate-officer-after-video-shows/

Glanfield, F., Thom, J., & Ghostkeeper, E. (2020). Living landscapes, archi-text-ures, and land-guaging algo-rhythms. *Canadian Journal of Science, Mathematics and Technology Education, 20*(2), 246–263.

Grosfoguel, R. (2011). Decolonizing post-colonial studies and paradigms of political economy: Transmodernity, decolonial thinking, and global coloniality. *Transmodernity: Journal of Peripheral Cultural Production of the Luso-Hispanic world, 1*(1), 1–38. https://escholarship.org/uc/item/21k6t3fq

Hogan, L. (2000). A different yield. In M. Battiste (Ed.), Reclaiming Indigenous voice and vision (pp. 115–123). Vancouver, BC: UBC Press.

Kawagley, A. O., & Barhardt, R. (1999). Education Indigenous to place: Western science meets Native reality. In G. A. Smith & D. R. Williams (Eds.), *Ecological education in action: On weaving education, culture and the environment* (pp. 117–142). Albany, NY: State University of New York Press.

Keene, A. (2018, February 24). Wakanada forever: Using Indigenous futurisms to survive the present. Native Appropriations. http://nativeappropriations.com/2018/02/wakanda-forever-using-indigenous-futurisms-to-survive-the-present.html

Kennedy, M. (2020, February 21). Why are swarms of locusts wreaking havoc in East Africa? NPR News. https://www.npr.org/2020/02/21/807483297/why-are-swarms-of-locusts-wreaking-havoc-in-east-africa

Kimmerer, R. W. (2021, January 29). A conversation with Dr. Robin Wall Kimmerer> University of British Columbia. https://register.gotowebinar.com/recording/846424254931495695

Kimmerer, R. W. (2013). *Braiding sweetgrass: Indigenous wisdom, scientific knowledge, and the teachings of plants*. Corvallis,OR: Milkweed Editions.

King, H. (2019, January 18). "I regret it": Hayden King on writing Ryerson University's territorial acknowledgement. *CBC Radio*. https://www.cbc.ca/radio/unreserved/redrawing-the-lines-1.4973363/i-regret-it-hayden-king-on-writing-ryerson-university-s-territorial-acknowledgement-1.4973371

Klein, N. (2007). *The shock doctrine: The rise of disaster capitalism*. Totonto: Vintage Canada.

Little Bear, L. (2000). Jagged world views colliding. In M. Battiste (Ed.), *Reclaiming indigenous voice and vision* (pp. 75–85). Vancouver, BC: UBC Press.

Lopez-Martinez, M. (2020, August 29). Protestors across Canada march to defund the police. *CTV News*. https://www.ctvnews.ca/canada/protesters-across-canada-march-to-defund-the-police-1.5084956

Lunney Borden, L., & Wiseman, D. (2016). Considerations from places where Indigenous and Western ways of knowing, being, and doing circulate together: STEM as artifact of teaching and learning. *Canadian Journal of Science, Mathematics and Technology Education, 16*(2), 140–152. DOI: 10.1080/14926156.2016.1166292

MacIvor, M. (1995). Redefining science education for Aboriginal students. In M. Battiste & J. Barman (Eds.), *First Nations education in Canada: The circle unfolds* (pp. 72–98). Vancouver, BC: UBC Press.

Manuel, A. & Derrickson, R. M. (2015). *Unsettling Canada: A national wake-up call.* Between the Lines.

Martens, K. (2020, June 4). Indigenous woman shot and killed by NB police. APTN National News. https://www.aptnnews.ca/national-news/indigenous-woman-shot-and-killed-by-n-b-police/

McMillan, L. J., & Prosper, K. (2016). Remobilizing Netukulimk: Indigenous cultural and spiritual connections with resource stewardship and fisheries management in Atlantic Canada. *Reviews in Fish Biology and Fisheries, 26*(4), 629–647.

Michell, H. (2005). Nēhīthâwâk of Reindeer Lake, Canada: Worldview, epistemology and relationships with the natural world. *Australian Journal of Indigenous Education, 34*(1), 33–43.

Palmater, P. (2014). Genocide, Indian policy, and elimination of Indians in Canada. *Aboriginal Policy Studies, 3*(3), 27–54. http://dx.doi.org/10.5663/aps.v3i3.22225

Paul, J. J., Lunney Borden, L., Orr, J., Orr, T., & Tompkins, J. (2019). Mi'kmaw Kina'matnewey and Mi'kmaw control over Mi'kmaw education: Using the master's tools to dismantle the master's house? In E. A. McKinley, & L. T. Smith (Eds.), *Handbook of indigenous education.* Springer https://doi.org/10.1007/978-981-10-3899-0_32

Prosper, K., McMillan, L. J., Davis, A. A., & Moffitt, M. (2011). Returning to Netukulimk: Mi'kmaq cultural and spiritual connections with resource stewardship and self-governance. *International Indigenous Policy Journal, 2*(4), 7.

Pueyo, T. (2020, March 19). Coronavirus: The hammer and the dance: What the next 18 months can look like, if leaders buy us time. Medium.com. https://medium.com/@tomaspueyo/coronavirus-the-hammer-and-the-dance-be9337092b56

Surkan, J. (2018). Conversations: Elmer Ghostkeeper. https://metisarchitect.com/2018/06/04/conversations-elmer-ghostkeeper/

Sutherland, D., & Henning, D. (2009). Ininiwi-Kiskanitamowin: A framework for long-term science education. *Canadian Journal of Science, Mathematics and Technology Education, 9*(3), 173–190.

Sylliboy, S. (2020). Valuing Mi'kmaw knowledge in the Nova Scotia science curriculum. [Unpublished master's thesis]. St. Francis Xavier University.

Thompson, D. (2020, May 6). What's behind South Korea's COVID-19 exceptionalism? The Atlantic. https://www.theatlantic.com/ideas/archive/2020/05/whats-south-koreas-secret/611215/

Truth and Reconciliation Commission of Canada. (2015). *Canada's residential schools: Reconciliation* (Vol. 6). http://www.myrobust.com/websites/trcinstitution/File/Reports/Volume_6_Reconciliation_English_Web.pdf

Tsing, A. L. (2015). *The mushroom at the end of the world: On the possibility of life in capitalist ruins.* Princeton, NJ: Princeton University Press.

Tuck, E., & McKenzie, M. (2015). *Place in research: Theory, methodology, and methods.* New York, NY: Routledge.

Unama'ki Institute of Natural Resources. (2016). *Netukulimk.* https://www.uinr.ca/programs/Netukulimk/

Valiante, G. (2020, April 9). Quebec preparing plan to reopen economy, as infection rate stabilizes. *The Globe and Mail.* https://www.theglobeandmail.com/canada/article-quebec-preparing-plan-to-reopen-economy-as-infection-rate-stabilizes/

Watt-Cloutier, S. (2021, February 9). From local to global [Online keynote]. Leadership Committee on English Education in Québec (LCEEQ) Annual Conference. https://lceeq.ca/en/events/qrcode/TZOQbRKDW2dRmtn0Q04jK839E0wY8Ey8Fzy6aZtQZzY

Wente, J. (2020, February 25). Reconciliation is dead and it never really was alive. *CBC News*. https://www.cbc.ca/news/canada/toronto/jesse-wente-metro-morning-blockades-indigen ous-1.5475492

Werner, J., & Lyons, S. (2020, March 4). The size of Australia's bush fire crisis captured in five big numbers. *ABC News*. https://www.abc.net.au/news/science/2020-03-05/bushfire-cri sis-five-big-numbers/12007716

Wilson, S. (2001). Editorial: Self-as-relationship in Indigenous research. *Canadian Journal of Native Education*, 25(2), 91–92.

Wiseman, D., Lunney Borden, L., Beattie, L., Jao, L., & Carter, E. (2020). STEM emerging from and contributing to community. *Canadian Journal of Science, Mathematics and Technology Education*, *20*(2), 264–280. https://link-springer-com.proxy.ubishops.ca:2443/article/10.1007/ s42330-020-00079-6

Wiseman, D., Onuczko, T., & Glanfield, F. (2015). Resilience and hope in the garden: Intercropping Aboriginal and Western ways of knowing to inquire into teaching and learning science. In H. Smits & R. Naqvi (Eds)., *Framing peace: Thinking about and enacting curriculum as "radical hope"* (pp. 237–252). Peter Lang Publishing.

Wiseman, D. (2018). Finding a place at home: The TRC as a means of (r)evolution in pre-service (science) teacher education. *McGill Journal of Education*, *53*(2), 331–349. http://mje.mcgill.ca/ article/view/9514/7358

Wiseman, D. (2016). *Acts of living with: Being, doing, and coming to understand Indigenous perspectives alongside science curricula* [Unpublished doctoral dissertation]. University of Alberta. https:// era.library.ualberta.ca/files/ctt44pm87m/Wiseman_Dawn_201603_PhD.pdf

Witze, A. (2020, September 10). The Arctic is burning like never before – And that's bad news for climate change. *Nature*, *585*, 33–337. https://www.nature.com/articles/d41586-020-02568-y

York, G., Chase, S., & MacKinnon, M. (2020, January 10). Iran admits 'disastrous mistake': Its military accidentally shot down Flight 752. *The Globe and Mail*. https://www.theglobeandm ail.com/world/article-iran-rejects-missile-theory-calls-on-west-to-show-evidence/

Zinga, D., & Styres, S. (2011). Pedagogy of the land: Tensions, challenges, and contradictions. *First Nations Perspectives*, *4*, 59–83. http://mfnerc.org/resources/fnp/volume-4-2011/

Land-Based

ndinawe-maa-ganag,
All our relations.
Lived with,
Were part of,
Lived in harmony - physically, mentally, emotionally and spiritually engaged.
lack of harmony,
        dis-ease,
            climate change,
                the SDGs,
                    MOOCs,
                        research and policy priority,
                        Fit-for-purpose.

                    Truth?

            Is this the time to shoot-through?
Share Kitchen, Foundational support.
Comparing traditional and western knowledge
Looking back and remembering
Being lost in time and being in that moment
Storytelling

passed on in a natural and all-encompassing way

Bush, Lakes, Forest.

Babies,

Future

Sacred responsibility

Wild bergamot, Berries, Medicine.

Stewards

ndinawe-maa-ganag

Reciprocity

Land,

Home.

Honoured.

CHAPTER FIVE

Anishinaabe Kwek Piimachiiwin: Indigenous Women's Anishinaabe Knowledge Systems

MYRLE BALLARD

INTRODUCTION

My name is Myrle Ballard. I am Anishinaabe ikwe from Lake St. Martin First Nation. I grew up speaking my language, my mother tongue, *Anishinaabe mowin*. I did not learn to speak English until I started school in grade 1. I grew up on the reserve and left there after I graduated from high school, to attend University. I used to complain about my childhood because I was expected to work [hard] and do chores, everything from cleaning house, to seaming fish nets, feeding cows, chopping wood, plucking ducks and geese, learning what to look for in the texture of the snow which was melted for our drinking water, picking rice and sugar beeting, to name a few. I listened but did not really pay attention to the stories that were shared with me: such as how the colour of sky when the sun sets forecasts the weather or how the leaves changed colour or how the sound of the rustling of the trees just before it rained. I did not realize until much later that these were *piimachiiwin*, the teachings that Anishinaabe kwek (women) passed on to children.

My interest in science started when I was a child. I grew up in my grandfather's house that was later passed on to my mother. This house was situated near the lake. By near, I mean less than a kilometre from the lake with an unobstructed view. I would walk to the lake during the summer to go swimming. Growing up I started noticing that "my" view of the lake was changing. The land between our house and the lake was starting to flood every spring/summer, and the vegetation was changing from hayland to swamp. While pursuing my post-secondary degree, I started to realize how important my *piimachiiwin* teachings were to my view of science and how they shaped who I am and where I am today.

This chapter is about the *piimachiiwin* teachings that were passed on to me through Anishinaabe mowin (language) by my mother, aunties, and the Elders–the Anishinaabe kwek –who shared stories with me. This chapter documents stories narrated by the Elders from Lake St. Martin First Nation (LSMFN) who were asked to share their knowledge, teachings, how they were raised, and the chores they did. The storytelling provided a rich narrative of how Anishinaabe teachings are transmitted through and embedded in Anishinaabe mowin. Particularly, I focus on Anishinaabe Kwek (Women)'s activities and knowledge. Anishinaabe kwek are the primary transmitters of Anishinaabe Knowledge Systems (AKS). The reason for this is because Anishinaabe kwek tend to be the ones looking after and raising children, but men also assist. Anishinaabe kwek–the mothers–are the primary caregivers, which also includes the transmission of the Anishinaabe mowin to the child, thus the term "mother tongue". The importance of this role of the transmission of Anishinaabe mowin cannot be stressed enough. The child learns the roles and learned behaviours of raising children, respecting nature, the land, and associated Indigen-ethics, and looking after family, through Anishinaabe mowin. The narratives provide a rich knowledge base of science, technology, arts, engineering, and mathematics, which the Anishinaabe kwek of LSMFN identify as Anishinaabe kwek pedagogy.

INTRODUCTION OF ELDERS

Throughout the years from 1998, 1999, and 2007 to 2009, I interviewed Elders (both female and male) from the First Nation communities of Lake St. Martin. These women included Elders Ruth Beardy, Roselyn Beardy, Edith Gagen, Priscilla Sinclair, Rosemary Forbes, Bertha Sumner, Evangalene Traverse, Flora Traverse, Rita Sinclair, and Grace Gabriel. The men included Elders David Ross and Mark Traverse. I knew most of the women and men by name. I invited some of them, including the two men, to attend three talking circles that were held in Lake St. Martin First Nation. I also conducted personal visits with some of them. The purpose of the talking circles and personal visits was for my graduate studies: M.Sc. on the topic of comparing traditional and western knowledge of water management and Ph.D. research on the role of gender and language in Anishinaabe knowledge systems. I reached out to them as these individuals all lived in the community and bring a depth of lived experience and knowledges. I wanted to learn about *piimachiiwin*. The Elders described *piimachiiwin* as the ways in which people survived and lived in the past. The lived experience is *piimachiiwin*, which are stories and teachings that are orally transmitted; stories that are important. *Piimachiiwin* transmitted by Anishinaabe Kwek (women) are about culture, traditions, food preparation, and livelihoods. Most importantly, piimachiiwin are shared in Anishinaabe mowin. These are called Anishinaabe knowledge systems (AKS) of Anishinaabe kwek.

ANISHINAABE KNOWLEDGE SYSTEMS: WE LIVED ON THE LAND!

Looking back and remembering the stories that were shared with me, I get a sense of being lost in time and being in that moment with the Anishinaabe kwek. Anishinabek normally represent all their teachings in a circle since the rhythm of life is cyclical (Mann, 2005). This principle is expressed in our languages–Anishinaabe mowin–as "all our relations." In Anishinaabe, "ndinawe-maa-ganag" means "all my relations," which is how the women/men of LSMFN also express themselves.

Anishinabek practiced sustainable management and were recycling long before the words "sustainable" and "recycle" were coined. When Anishinabek lived in the bush for weeks at a time they did not destroy or damage the land. Anishinabek did not have any disposable or plastic material with them, so they did not leave garbage when they left. The little waste they had was burned. The Anishinabek were very capable stewards of their land because they lived on the land. Tampering, tainting, contaminating, and /or polluting is not part of the way the people of the earth are supposed to live. The Elders I spoke to said that God assigned us to be stewards of this land. This understanding is what makes us Indigenous peoples.

Elder Ross (1999) told me that Anishinaabe Knowledge Systems and the land with all its resources are seen as a gift from Muntoo. Entire communities are dependent on ancestral territories for the survival and well-being of everyone within the group. The focus here is on the entire group. The land provides every-thing. For the Anishinabek *uundatisii* is significant because it means "living off." For Anishinaabe land is called *du kii naan* meaning "our land". The knowledge of the land, enabled the Anishinabek *piimachiiwin*–"they knew how to survive and live off the land." Anishinabek *kii pi majii yuu* meant that the Anishinabek as a community were able to survive by looking after themselves and each other, now known as livelihoods. This meant that Anishinabek were able to live off the land sustainably and what it had to offer. *Uundatisii* is a concept or word that is held with high regard by the Anishinabek. This also refers to *uundatisii* which means "being able to live off" or "being sustained by something."

The differences between Western and The Anishinabek traditional knowledge systems are contrasted by the difference in the lifestyle of neeji Anishinebek prior to the availability of "modern conveniences." The Anishinabek talk about the living earth, the living water, the living wind, and the living fire, everything is living. A nomadic existence necessitated the need to seek new resources and to allow for the natural replenishment of those left behind. Elder Flora Traverse (1998) shares, when Lake St. Martin First Nation flooded in the late 1800s, the Anishinabek moved to the south side of the Narrows to allow for the rejuvenation of the land. When the Anishinabek arrived at their destination, they would scout the area for water. Elder Gagen (2007) remembers, "They would look in the tall grass or even in the muskeg and that is how they found the water. The water was good. There was no pollution at that time. The water was in the grass, even a little water." Drinking water was obtained by digging a hole in the ground that would cause the natural water to come up. The water–*muskego wabo*–or water from the ground, was brown in color, but very cold and good. "We never even boiled the water; we just drank it straight from the ground" (Elder Flora Traverse, 2007). The Anishinabek never boiled their water and were never under any boil water advisories unlike today where the water on their land is contaminated. Tagore (2004) states that water does not merely cleanse the limbs but also it purifies the heart, for it touches the soul. The earth does not merely hold its body but also it gladdens its mind, for its contact is more than a physical contact–it is a living presence. The Elders spoke of how no one got sick when they spent weeks, even months at a time living in the bush. Elder Gagen (2007) remembers how they used to make their beds on the ground with only a canvas or hide separating them from the ground. She shared that "it's this closeness laying next to ground that builds that bond with the land." The closeness to the earth was in fact the physical closeness.

Anishinabek lived with nature, were part of nature, and lived in harmony with their land. This harmonious relationship was conducive to their well-being and did not make them sick. The relationship was reciprocal and close. Anishinabek only took what was needed and they did not pollute it. Einhorn (2000) mentions the reciprocity by stating that "each part of creation tries to form a physically and spiritually harmonious relationship with all other parts. Harmony and balance equal survival and happiness. Lack of harmony and balance causes sickness and death or, in short, dis-ease" (p. 22). In Anishinaabe mowin, the words are powerful when they describe the state of one's physical state. The physical harmony of "total well-being" is referred to as *mino-ayawin*. The opposite of well-being is *mazi-ayawin* which refers to not being well. *Mino* means "well" and *aya* means the state of being, *mino-aya* combined means "well-being." *Mazi* refers to not being well. During an Elders Gathering for healing, Elders referred to *mino-ayawin* as a clean and safe environment which is linked to the environmental determinants of health which are clean water, clean air, living off the land, access to traditional foods, being on the land, living in harmony with nature and a positive relationship with everything that includes people, animals, and all living thing (Ballard & Martin, 2017).

ANISHINAABE KWEK PEDAGOGY

The Anishinaabe kwek acknowledge that women are very strong because they are the child bearers. Women were considered the source of human life. Their role as life bearers and mothers was respected and acknowledged. Today, in traditional ceremonies, men and women always reference women as being the source of life. According to the Haida Gwaii (1989), the significance of women is premised on natural law: The Natural Law gives women the responsibility of bringing new life into the world. Everyone must be born from the womb. Mothers must protect the lives they have helped to bring into this world. The Anishinaabe kwek have respect for their purpose in life as child bearers. When they talk about women's "monthlies" they refer to it as *kaygo kaiziayad ikwe*–"when the woman is going through something". The English word for "menstruation" is not revealed in Anishinaabe mowin. The "monthlies" are sacred and considered powerful. Even the term *kaygo kaiziayad ikwe* does not translate to the word "menstruation" that is used in English. It refers to the sacredness of the time of month a woman goes through. The women also take precautions during "that time" of the month not to get cold and for women to take care of themselves. The Anishinaabe kwek teach women of childbearing years to take care of themselves during menstruation *kaygo tackuch iicane*–look after themselves and "not to catch [a chill]." The meaning–*kaygo tackuch iicane* cannot be fully translated but it is a very powerful teaching that is passed on.

The transmission of a knowledge system and Indigenous pedagogy through Indigenous languages is recognized as an important part of one's identity and is integral to learning. Anishinaabe language or Anishinaabe mowin is the principal instrument by which Anishinaabe kwek (women) transmit their knowledge from one generation to another. Because Anishinaabe mowin translates the world and experience in cultural terms, Anishinaabe mowin literally shapes our way of perceiving—our worldview. The Anishinaabe worldview has long been and still is shaped by life close to the land and a deep appreciation of the spiritual dimension of being.

However, in traditional land-use studies, traditional ecological knowledge (TEK) studies and in studies of Indigenous Knowledge Systems (IKS) women's knowledge is largely absent. A Cree Scholar from northern Saskatchewan, Priscilla Settee (2007) raises awareness about the oversight of women's large role in the transmission of knowledge and this focus on hunting, trapping, and fishing activities, which are conducted almost exclusively by men in today's world of the Anishinabek. The lack of studies has resulted in a gap regarding women's knowledge and Anishinaabe kwek pedagogy. Traditional gathering activities such as berry picking, crafting, and medicinal plant gathering are gender-biased. The gathering activities have been romanticized to be predominately Anishinaabe kwek activities, and naturally the typical western researcher on TEK was the white male. Therefore, the gathering activities are not considered as a priority for academic investigation. Most of the women's gathering takes place mostly on reserve or close to the reserve and not in areas that are of interest for industrial or conservation purposes.

In a similar way regarding how woman's knowledge has been absent in studies of IKS, so too has the voice and knowledge of women been historically absent in STEM. It has been noted by researchers and policy makers that historically the STEM field (science, technology, engineering, and math) have been male dominated. This gender disparity continues today. Wells (2019) found that women made up 44% of undergraduate students in STEM fields in 2010, compared to 64% of non-STEM fields. This gender disparity has been also seen in the results of STEM research in fields such as medicine. Many diseases and drugs have been studied in men only with women being considered "small men" (Glezerman 2009), but in reality, men and women respond differently in both physical and mental health fields (CIHR, 2012). Similar to the disparity of men's vs. women's knowledge in IKS, this mindset of valuing men's knowledge and being in STEM has resulted in less funding available for education, research and studies about women, for women and by women.

This holds even more true when considering the voice of Indigenous women in STEM. The intersection between discrepancy of women in STEM and Indigenous Peoples in STEM creates an environment where knowledge systems, research, natural law, policy, etc. are missing the unique voice and ideas that Indigenous Peoples, especially women, can bring. Over time a lot of the research and analysis about IKS has been provided by non-Indigenous men. The impacts of intersectionality have recently been acknowledged by the Canadian Government's Impact Assessment Act that specifies Gender Based Analysis Plus (GBA+) must take into account environmental and social considerations that affect people of vulnerable ethnicities, age, gender, etc. in different ways. GBA+ also recognizes that due to the historical and current dominant society laws, policies, etc. that people who intersect in vulnerable categories will be disproportionately affected (Canada Impact Assessment, 2020). Indigenous women's knowledge and application of that knowledge intersects with the Crown (government) system of administration, with analysis being mostly from men. Non-Indigenous women's knowledge and application of their acquired knowledge of the natural world through science, technology, engineering, and mathematics intersects with an academic and professional culture that is dominated by men.

The accumulated knowledge and understanding, including that of ecological and spiritual, are crucial for Anishinaabe kwek. The women of LSMFN have definitions, values, and systems of classification in place, although they are not formalized on paper. This whole chapter discusses the foundations of AKS with its interrelationship with native spirituality as a context to consider the role of women in AKS over time from the 1930s to more recent times from the viewpoint of the Elders that were interviewed. The Elders I spoke to defined Anishinaabe Knowledge Systems as to the systems of knowledges from hunting, gathering, trapping, fishing, looking out for and after the ecosystem, being the voice for the flora and fauna, and how everything works together and in harmony. The natural laws of the ecosystems have ways to balance things out so that the natural balance is always restored for life and survival for all species guaranteed. Anishinabek understand this as *aniin kaygo ayzi anokiimaguk*–or "how things work".

In LSMFN, the women balanced their lives between living in the bush with their families for weeks at a time in the summer with their homes on the reserve at other times. Bush living was a necessity for food on the trap line and not a retreat for peace and harmony, although it did provide that. However, the women recall that this life was not harsh but rewarding and the only life they knew. People learned by working on the land. In the next section, I share some of the Anishinaabe Kwek's activities and knowledges that were shared and practiced for thousands of years from working on the land and water.

A Women's House: Task is Knowledge!

Anishinaabe kwek were very diligent in their roles to maintain a healthy home environment. Children, when able, helped with all the necessary chores. There were no washing machines and all laundry was washed by hand. Elder Evangalene Traverse (2007) remembers how they used to make a day trip to the lakeshore to do their laundry. They would stand in the water and wash their clothes and hang them on the trees along the shore to dry. They made their own soap from moose fat, which was mixed with ash from the fire.

All tasks in the home to improve life were done by women. Mattresses were made by stuffing straw into canvas material (Elder Forbes, 2009) and changed at least once a year by replacing the straw. Pillows were made from the scraped moose fur and stuffing it into canvas (Elder Forbes, 2009). The Anishinaabe kwek were also strong and provided much leadership as well as raising children. Life is easier today compared to the days when women would look after big families and everything would be washed by hand including diapers. Today, Anishinaabe kwek would not be able to cope if they had to use diapers and wash them every day (Elder Sumner, 2007).

Fall Harvest: Enough to Last Them All Winter

After the August harvest moon, the Anishinaabe kwek prepared for the fall by gathering roots and medicines. These roots and medicines were important to keep the people healthy during the long winter months. They were made into *muskiki* or *muskiki wabo* (explained later on). The Anishinabek knew what the weather and conditions would be like in the coming seasons and prepared themselves in anticipation. *Weekay* was and is still a popular medicinal plant that has many purposes. It was gathered at a certain time and prepared for storage to be used during the winter months. Elder Beardy (2009) remembers how their parents only harvested enough *weekay* to last them all winter, which was about two inches of root. There were also numerous other medicinal plants that were harvested and stored for use during the winter months.

The Anishinabek knew when to harvest plants. Western science recognizes the chemical changes that occur within plant roots as the seasons change but for the Anishinabek, it is more than recognizing the physical changes in the plant. The Anishinabek take note of the changes in the weather, how the winds blow, how the birds fly, the colour of the sunset, the colour of the sky, the colour of the halo around the winter sun, how the air smells, how the birds and animals behave, how early or how late the plants blossomed, and the list goes on (Elders, 2009). In the summer, certain plants are gathered at solstice time, when the sun energy is the longest and the highest. This induces the productions of certain healing qualities in the plants that make them more powerful for healing certain ailments. After being gathered at solstice they were dried in the open to receive even more sunlight.

The Anishinaabe kwek used various medicines, which were made from plants. They had medicines for everything. When a child had diaper rash the bark from a certain tree was used. It was finely chopped until it looked and felt like powder. "It looked like the baby powder that is used today; that's how good it was and the rash healed right away no matter how bad the rash was. Back then we had no pampers just diapers" (Elder Sumner, 2009). They also used moss on babies (Elder Sumner, 2009). When babies had rashes around their necks they used *piwaskinag* which are the little plants that hang in bunches. They would gather the *piwaskinag* and when they turned into powder they used that on children for neck rash and diaper rash (Elder Sumner, 2009). Rabbit skin was used as well for covering babies. "We got all of our medicines and food from nature and we never got sick" (Elder Gagen, 2009).

Knowing the Elements and How to Nurture Healthy Babies

The women always took extra precautions during the winter against nature's elements. They knew the force of nature and respected its force. When they took infants and children outside in the winter they made sure that they were properly sheltered and covered from the cold wind and cold air. They wanted to prevent the child from *keewatinownin uku pootanigoon*–"the north wind will blow on him". The women know that there are certain elements in the winter air that can make children sick or even kill them. Also, during spring as the snow and ice thawed, the Anishinaabe kwek took extra precautions to prevent the child from *miicomiin uku pootanigoon*–"the ice will blow in his face". The translation or meaning has no relevance in English, however, in Anishinaabe, it is a powerful warning for the mothers to take extra precautions for her baby during that time of the year. Another warning was *kaygo tudokujchid*–"not to catch a chill".

Traditional healers were consulted about the health of children. Even today, the Anishinaabe kwek who speak Anishinaabe mowin know "the child that gets sick from cold teeth". One of the Anishinaabe kwek, Gabriel (2008) recalls taking her grandchild to a traditional healer for inflamed gums. The traditional healer made *muskiki wabo* (muskiki is the medicine, wabo is the medicine as a liquid) from plants and put it into a cheese cloth. The little bag was filled with plants and was rubbed inside her grandchild's mouth. The traditional healer showed her what to do so that she could treat her grandchild at home and administer the *muskiki* herself and provide her with this medicine for the baby. When drinking the *muskiki wabo*, the mother was advised that the baby should not drink any type of milk while on the medicine. A sweetener could be used to take away the bitter taste such as sugar or syrup. The women from LSMFN also talked about this *muskiki* and how it is made. The plant was picked from an open field. The plants are in the shape of a little ball with plant fluff inside them. The plant was prepared and used specifically for *katuka tipaywujchid abinoonjii*– "the child who gets infected from cold teeth" (Elder Sumner, 2009). The term is a condition in infants that does not have a medical term in the English language, the closest would be "teething" but the translations get lost and does not mean the same thing. *Katuka tipaywujchid abinoonjii* cannot be treated by western medicine like antibiotics, and is said to make the infant weaker, so Anishinaabe mothers still seek traditional medicine for their children.

The Berries Are Not as Abundant

The Anishinaabe kwek were the primary gatherers. Men were also gatherers however it was the women who took the lead in initiating the gathering outings. The Elders talk about how they spent days picking different berries and getting ready for the winter. The berries were picked according to when they were ready for picking. It was usually the women who kept track of the seasonal patterns regarding when the berries would be ready. The weather patterns were particularly important when they were preparing for the winter. For example, during the late spring and the season leading into early summer, the women know if there are going to be certain berries that will grow abundantly. The Anishinabek knew when to go out. Sadly though, during the past decade, the Elders (2009) have noticed that the berries are not as abundant as they used to be. "We all used to go picking berries–low bush cranberries–*muskiigiminim*–with our parents. It used to look like the berries were just spilled over–that's how abundant they were and they all used to be so delicious. We would make jam and have jam all winter long" (Elder Sinclair, 2009).

There used to be a lot of blueberries by Big Rock every summer and yet today there are none around. Saskatoons were also picked and were in abundance, as well as strawberries. Elders Gagen and Traverse (2009) shared how they went blueberry picking at Big Rock and go there by horse and wagon and were gone for 2 to 3 days. "We would bring back 4 to 6 tubs. We used fish wooden tubs that were about 2½ feet wide. Many people used to go blueberry picking. They were not small either, but big–today there's not that many around and they're small. We had long hot summers in those days, not like the summers today" (Elder Traverse, 2009).

Most of these berries also have Anishinaabe names indicating that they have been a diet staple, namely: *nthigokuminuk* (Saskatoon berries), *aniibiminum* (cranberries), *mithkuminuk* (raspberries), *miinun* (blueberries), *udayminun* (strawberries), and *muskiigiminun* (low bush cranberries). The berries were also prepared and preserved in different methods depending on their use. Saskatoons were made into jam for use in the winter or dried [like raisins] (Elders 2009). Produce from gardens was also preserved. They did the same with corn and onion (Elders 2009). Corn was dried for use during the winter. Potatoes were grown in gardens and kept well during the winter months in holes dug in the ground. The women were the ones who tended the gardens, but it was the men who prepared the gardens by tilling them. The Anishinabek used to keep everything they picked–*mawinzowin*–and would keep it all winter (Elders 2009).

Nothing Was Wasted: Harvesting and Preparing Meat

When families went hunting, they would be gone for about a week or until they had enough meat (Elders 2009); they stayed until they killed a moose. Elder Sumner (2009) from Lake St. Martin First Nation reminisced of her younger days, "My dad used to kill moose and we used to tan the hide to make moccasins … They used to smear the moose brain on the hide and it used to soften it. It would stay on until it was nice and soft. After it softened, they twirled it around with branches. Then it was pulled and dried by the fire. They scraped it further to make it softer and cleaner … They used spruce which was rotten—it's a specific type that looks like ash—powdery. This was also used for diaper rash. They also made muskeg as a powder to prolong the dryness of a diaper. Moose hair was used in the winter as bed fill. We also used hay as bed fill. Nothing was wasted." They prepared and dried the meat while they were at camp (Elders 2009). When a moose or bear was killed, they made lard—*thathkiganang*—from the fat (Elder Gagen, 2009). Moose fat was also used to make soup, to make bannock and pastry such as pies. Bear fat tasted better than moose fat. *Thathkiganang* had many uses, one if which was to make lamps. This was used instead of coal oil, also used to light homes at night, but many people could not afford it. They used to take a piece of rag and twist and wrap it around, grease it with lard from moose or bear fat, put it in a saucer, light it, and it provided a dim light. This was called a skunk lamp.

Every part of the meat was used, and none was wasted. The moose used was not only for consumption but served many purposes. The Anishinaabe kwek never used thread. They used *sinew* as a thread which is the muscle found along the back of moose and along the loins and is used when it dries up. They also used *udith* as thread, which is from the moose belly button. These were very strong and did not break and did not rip even when you tried ripping it. When they made moose hide shoes and moccasins that is what they used. They also used them for violin strings, using different thickness for each string.

Fish was also a staple in the Anishinaabe diet. The men fished commercially to make a living as well as providing food for their families. However, the women also fished, especially when the men were not home, as a means of sustenance. The women also prepared the fish for consumption and storing. Elder Traverse (2008) remembered how she used to go fishing with her mother during the summer when her father was away working. "My mother and I would go out on the boat and she would throw a net into the water to catch fish for us to eat. We caught enough fish to last us for a few days. When we got home my mother prepared the fish. She used all parts of the fish and did not waste anything. I remember her frying the fish guts with onion and I can still remember that smell. It smelled good and tasted so good" (Elder Traverse, 2008). *Muunuyigaing,* digging for roots, is a skill that is learned, and the majority of the time, it is the women who teach the children and youth how to dig for roots. The women would show what the root looks like and what the colour should be, when is the best time to dig, and how to dig properly. When the Anishinabe kwek dig for root they do it very fast moving from place to place. They also know what type of vegetation it grows best in and where to find it. In recent years, digging for Seneca root is done to supplement income, and for some it is also a chance to go back into the bush and spend the day there.

Sewing and the Art of Preparing Hide

Preparing hide was not an easy task. Sticks were used to tie the hide in place and a fire was kept burning under it to smoke the hide. Two people had to stand pulling the finished hide back and forth over the fire for at least half a day. After the curing process the hide was nice and brown and it was light like paper. When the hide was good and dry, they used it for sewing and making different items like slippers and wrap-arounds for children (Elder Beardy, 2009).

Sewing and beading were important skills. Not everyone could bead. The women that did bead were very good at what they did. In early times, all clothes were made out of hides and then later cloth. In the wintertime the women made jackets and pants for the men. The women made parkas with white canvas cloth. They also made men leather pants made from moose skin. Women knew how to sew and all by hand because no one had machines (Elder Sinclair, 2009).

Families Camped on the Edges of Hayfields: Agriculture

First Nation people engaged in agriculture long before settlers. Archaeological digs found the *three sisters*—corn, beans and squash—have been found throughout the Americas long before settlers arrived (LaDuke, 2005). Agriculture was an important food source traditionally, along with fish and meat. The importance of agricultural activities during the 1930s has been reported for other Aboriginal peoples in Manitoba and Saskatchewan as well (Buckley, 1993). The Anishinabek at LSMFN were horticulturalists, agriculturalists, and had cattle from as early as 1900s.

Potato gardening and haying filled out periods that were of low activity seasons for traditional undertakings. Making hay became an additional seasonal activity of great economic and social importance. From late July on, entire families would come together and camped on the edges of the hay fields. These kinds of haymaking camps lasted into the 1960s. Elder Traverse (2009) remembers how they used to walk across LSM narrows to go to the other side. "The other side (of LSMFN) had lots of hay and that's where we spent our summers and the hay was done" (Elder Traverse, 2009). Elders still talk about these haymaking times with great joy. Potatoes were mostly harvested at the beginning of October after the fall moose hunt.

Until recently, Anishinabek still practiced their "subsistence" livelihoods, but the focus is not the same. The Anishinabek were rice picking up until the 1980s, however modern conveniences were intermixed and also it was part of the commercial process. It shifted from subsistence to commercial.

Salt was also made for consumption. "There's a lot of natural salt around this area and they used to make it for table use. They also used it for animals to eat–the deer especially like the salt–*tamajiiskiiway*" (Elder Traverse, 2008). They looked for the raw salt by looking for white mounts, which was boiled dry to purify it (Elder Traverse, 2008).

Wiichiitwin: Building Houses Together

All the families that had horses gathered firewood for those families who did not have horses and unable to fell wood for themselves. *Wiichiitwin* meant a time of helping. "The families cooked together, ate together, and then when they were all done had a dance party without alcohol. In those days there was no alcohol" (Elder Gagen, 2009).

Kiimbeekaywug involved the whole community coming together to help someone. For example, if someone needed a house, the activity of coming together to build the house or help some in need was *kiimbeekaywug*. The term does not imply any form of building or construction but refers to the community coming together and helping one another. The Anishinabek looked after one another and those in need. This was a community fundraising–without money, where the people got together to help someone in need. Interestingly enough, the term is no longer used today.

The men typically did most of the heavy labour, but the women also had a role in the home building. The women prepared the food and gathered the wood for the fire. Unlike today where all the building material is pre-cut into lumber of various sizes and uses, the men would go into the bush and look for trees to be used of a specific type, height, age, and appearance. The men brought back the trees and prepared them into logs by manually cutting off the branches and cutting the trees into logs for building houses using axe and hand saw. The logs had to be cut so that the corners met and overlapped each other. The men also made the mud to fill the spaces between the logs. There was a lot of work involved in building the log houses. "The men would shove the mud in between the logs and then they [the houses] were painted white" (Elder M. Traverse, 2009). This paint was not store-bought paint, but the colour came from heating limestone. The men rotated with each other to keep the fires going all night to heat the stone. "All night long there was a man watching and all day long they worked hard" (Elder M. Traverse, 2009). They heated stones by digging big holes in the ground where they would cook the stone there it until it was white. Upon heating the limestone would break up into little pieces, and then was further crushed to look like powder. After the limestone was crushed, they mixed it with water to get a paint-like mixture, which was then used to paint or plaster the houses. "It was hard work transporting those stones. They got the limestone from all over. It was just laying around. They just went around collecting them" (Elder M. Traverse, 2009). But the stone was not just any limestone. It had to be a particular colour and it had to be a specific texture. "The stones were white and they were soft and they knew which stones to pick" (Elder M. Traverse, 2009). The limestone used to be hauled by the Anishinabek by horse and wagon. The houses were warm in the winter and much warmer than they are today. "There was no electricity. They used firewood all winter" (Elder M. Traverse, 2009).

IMPORTANCE OF ANISHINABE KWEK PEDAGOGY AND ITS ROLE IN EDUCATION

When I spoke to the Elders (2009), they shared stories with me of how the women sat on the ground, ate on the ground, and slept on the ground. They did not have chairs or tables in the bush, unlike today, where people go camping and they take tables, chairs, and air mattresses with them. On the land Anishinabek find a level of understanding and an all-pervading spirit that shapes them profoundly in their human development. Being on the land gives strength, health, and vitality that cannot be achieved with reserve living conditions or in cities. However, today, the deeper connection to the land, being in the "bush", is missing. When one does a lot of activities with direct contact with the ground from being on the land and being in the bush, a very special relationship starts to happen with the earth. Feelings of a strong grounding, safety, security, strength, and well-being start to emerge (Elder Gagen, 2009). One feels rejuvenated, refreshed, and balanced from longer contacts with the ground. One gains a different kind of respect towards the Earth from spending longer times on the ground. Generally speaking, this is something that is missing in people's lives today. The expression of people "being grounded" is commonly used to refer to well-being.

When people are working on the land, they are in a state of being present in the moment. They are physically, mentally, emotionally, and spiritually engaged in their activities. This is the time when traditional knowledge is passed on in a natural and all-encompassing way. This is when the words, which go along with the activities, make the most sense. There is a certain mental strength that is required to live a traditional lifestyle with bush activities on a regular basis. Experienced Elders and traditional land users pass on this strength to the younger inexperienced people so that they can become strong people of the land in order to look after their families. The experienced people make quiet suggestions of how somebody can improve their skills. With time the younger inexperienced people gain experience, confidence, and respect. They become more and more part of the team. With every step of earning their knowledge and respect, more information is then passed on. The Elders know when somebody is ready to receive the next level of knowledge. They call this *mi aza tu ma ji kakina mawiind*.

Incorporating STEM into everyday life and on-the-land teaching is key in involving Indigenous girls and women in STEM fields. One of the many ways of learning is that of hands-on learning. As children, we have a natural curiosity for observation of land, plants, and animals that is valued by many Indigenous communities (Sprang & Bang, 2014). Framing STEM in ways that are local to each community and each person, is also key to making STEM fields immersive in everyday life. These observations can be communicated to each other in culturally relevant ways, such as oral tradition and storytelling on their own as well as working together with technology (Sprang & Bang 2014; Lee et al., 2019).

When people become involved in any of the STEM fields, they do not do so in isolation, they bring their cultural values and worldviews with them (McKinley, 2016). Rather than seeing this as a negative impact to STEM, it brings new ideas and ways of knowing into STEM fields. Indeed, Western Science may accept Indigenous Science as new ideas, but rather, Indigenous Science has been transmitted since time immemorial through oral histories and traditions. Western science only arrived on Turtle Island by accident in 1492. Many researchers have recognized the role that culture plays in teaching, learning, and research in schools and beyond (Sprang & Bang 2014; Lee et al., 2019).

CONCLUSION

This research defined me and paved the path to where I am now. I learned to appreciate my mother tongue and hold my language—Anishinaabe mowin—in high regard. I started to understand that language has an important role in the transmission of knowledge, which goes largely unnoticed, and which is attributed to the fact that Anishinaabe kwek are the transmitters of language, and that their teachings and passing of their knowledge is not deemed as important by researchers.

There is a huge gap in literature on Anishinaabe women knowledge systems. In order to understand the dynamics of sustainable livelihoods and Anishinaabe women, more research should be conducted, including the lack of Anishinaabe speaking researchers has been a huge limitation in this field of research. Stories and teachings from the past need also to be incorporated in STEM.

Anishinaabe kwek have played an important role in the transmission of Anishinaabe Knowledge Systems (AKS). Women have shared this role with men, although both women and men have different roles at the community level. The

Anishinaabe kwek are the transmitters of AKS in terms of child rearing, food preparation, looking after their households and families. All of these roles played a significant part in Anishinaabe kwek role as stewards of the land. AKS guides spiritual, cultural, and social dynamics of the Anishinabek. The AKS is rich in painting a vivid picture of the subsistence economies of the Anishinabek, which were further supplemented by income, based on their knowledge of the land. The Anishinaabe kwek had an important role in the transmission of AKS, but this did not preclude the men from the responsibility of knowledge transmission. Each had a role to play, as individuals and as a community. With time and the gradual phase-in of technology, and the implementation of government policy upon their lives, AKS have eroded. Anishinaabe mowin which was once the base of a culture steeped in tradition is now on the verge of being lost as modern lifestyles and conveniences compete with the simplistic, yet physically demanding, traditional way of life. Indigenous women, in particular, are considered to be the stewards of the land. However, Indigenous women's role in resource stewardship is typically ignored in academic literature. This research found Indigenous women are still the backbone today as shared and told by the Elders in this research (Elders, 1998, 1999, 2007, 2008, & 2009). In conclusion, Anishinaabe women's traditional knowledge systems are very important. Many of us still carry on and remember the teachings from Anishinaabe kwek.

REFERENCES

Ballard, M., & Martin, D. (2017). *Hearing the elders' voices: Minoayawin*. University of Manitoba.

Buckley, H. (1993). *From wooden ploughs to welfare: Why Indian Policy failed in the Prairie Provinces.* Montreal & Kingston: McGill-Queen's University Press.

Canadian Institutes of Health Research (CIHR). (2012). *What a difference sex and gender make: A gender, health and sex research cookbook*. Retrieved from https://cihr-irsc.gc.ca/e/44734.html

Einhorn, L. J. (2000). *The Native American Oral Tradition*. Westport, CT: Praeger.

Elders. (1998). Elders, Lake St. Martin. Personal Communication. Lake St. Martin, MB.

Gagen, E., Lake St. Martin First Nation. Lives in Lake St. Martin, MB. Oral history, personal communication. August 4 & 5, 1998.

Sinclair, P., Lake St. Martin First Nation. Lives in Lake St. Martin, MB. Oral history, personal communication. August 4 & 5, 1998.

Sumner, B., Lake St. Martin First Nation. Lives in Lake St. Martin, MB. Oral history, personal communication. August 4 & 5, 1998.

Traverse, E., Lake St. Martin First Nation. Lives in Lake St. Martin, MB. Oral history, personal communication. August 4 & 5, 1998.

Traverse, F., Lake St. Martin First Nation. Lives in Lake St. Martin, MB. Oral history, personal communication. August 4 & 5, 1998.

Elders. (1999). Elders, Lake St. Martin. Personal Communication. Lake St. Martin, MB.

Gagen, E., Lake St. Martin First Nation. Lives in Lake St. Martin, MB. Oral history, personal communication. July 20 & 21, 1999.

Ross, D., Lake St. Martin First Nation. Lives in Lake St. Martin, MB. Oral history, personal communication. July 20 & 21, 1999

Sinclair, P., Lake St. Martin First Nation. Lives in Lake St. Martin, MB. Oral history, personal communication. July 20 & 21, 1999.

Sumner, B., Lake St. Martin First Nation. Lives in Lake St. Martin, MB. Oral history, personal communication. July 20 & 21, 1999.

Traverse, E., Lake St. Martin First Nation. Lives in Lake St. Martin, MB. Oral history, personal communication. July 20 & 21, 1999.

Traverse, F., Lake St. Martin First Nation. Lives in Lake St. Martin, MB. Oral history, personal communication. July 20 & 21, 1999.

Elders. 2007. Elders, Lake St. Martin. Personal Communication. Lake St. Martin, MB.

Gagen, E., Lake St. Martin First Nation. Lives in Lake St. Martin, MB. Oral history, personal communication. July 20 & 21, 2007.

Sumner, B., Lake St. Martin First Nation. Lives in Lake St. Martin, MB. Oral history, personal communication. July 20 & 21, 2007.

Traverse, E., Lake St. Martin First Nation. Lives in Lake St. Martin, MB. Oral history, personal communication. July 20 & 21, 2007.

Traverse, F., Lake St. Martin First Nation. Lives in Lake St. Martin, MB. Oral history, personal communication. July 20 & 21, 2007.

Elders. 2008. Elders, Lake St. Martin. Personal Communication. Lake St. Martin, MB.

Beardy, R., Lake St. Martin First Nation. Lives in Lake St. Martin, MB. Oral history, personal communication. July 24, 2008

Gabriel, G., Lake St. Martin First Nation. Lives in Skownan, MB. Oral history, personal communication. June, 7, 2008

Sinclair, R., Lake St. Martin First Nation. Lives in Lake St. Martin, MB. Oral history, personal communication. July 23, 2008

Traverse, E., Lake St. Martin First Nation. Lives in Lake St. Martin, MB. Oral history, personal communication. July 22 & 23, 2008

Elders. 2009. Elders, Lake St. Martin. Personal Communication. Lake St. Martin, MB.

Beardy, R., Lake St. Martin First Nation. Lives in Lake St. Martin, MB. Oral history, personal communication. August 27, 2009

Forbes, R., Lake St. Martin First Nation. Lives in Lake St. Martin, MB. Oral history, personal communication. August 25, 2009

Gagen, E., Lake St. Martin First Nation. Lives in Lake St. Martin, MB. Oral history, personal communication. August 24, 2009.

Sinclair, P., Lake St. Martin First Nation. Lives in Lake St. Martin, MB. Oral history, personal communication. August 21, 2009.

Sumner, B., Lake St. Martin First Nation. Lives in Lake St. Martin, MB. Oral history, personal communication. August 26, 2009.

Traverse, E., Lake St. Martin First Nation. Lives in Lake St. Martin, MB. Oral history, personal communication. August 26, 2009.

Traverse, M., Lake St. Martin First Nation. Lives in Lake St. Martin, MB. Oral history, personal communication. August 24, 2009.

Glezerman, M. (2009). Discrimination by good intention: Gender-based medicine. *The Israel Medical Association Journal: IMAJ* 11 (January). Retrieved from https://www.researchgate.net/prof ile/Marek_Glezerman/publication/24255454_Discrimination_by_Good_Intention_Gen der-Based_Medicine/links/00b4952bddde3d34de000000.pdf.

Haida Gwaii. (1989). Traditional Circle of Elders. Haida Gwaii. Queen Charlotte Islands Skidegate – Massett June 14, 1989, COMMUNIQUE NO. 12.

Impact Assessment Agency of Canada. (2020). Practitioner's Guide to Federal Impact Assessments under the Impact Assessment Act. Retrieved from https://www.canada.ca/en/impact-assessm ent-agency/services/policy-guidance/practitioners-guide-impact-assessment-act.html.

LaDuke, W. (2005). *Recovering the sacred: The power of naming and claiming.* Toronto: South End Press.

Lee, A. S., Wilson, W., Tibbitts, J., Gawboy, C., Meyer, A., Buck, W., Knutson-Kolodzne, J., & Pantalony, D. (2019). Celestial calendar-paintings and culture-based digital storytelling: cross-cultural, interdisciplinary, STEM/STEAM resources for authentic astronomy education engagement. *EPJ Web of Conferences, 200* (01002). https://doi.org/10.1051/epjc onf/201920001002

Mann, C. C. (2005). 1491: *New revelations of the Americas before Columbus.* New York: Alfred A. Knopf.

McKinley, E. (2016). STEM and indigenous students. In *ACER Research Conference Proceedings.* Camberwell: Australian Council for Educational Research.

Settee, P. (2007). Pimatisiwin: Indigenous knowledge systems, our time has come. Ph.D. Dissertation, University of Saskatchewan.

Sprang, M., & Bang, M. (2014). *STEM teaching tools implementing meaningful STEM education with indigenous students and families.* Retrieved from http://stemteachingtools.org/brief/11.

Tagore, R. (2004/1913). *Sadhana. the realization of life.* New York: Three Leaves Press.

Wall, K. (2019). Persistence and representation of women in STEM programs. *Insights on Canadian Society.* Retrieved from https://www150.statcan.gc.ca/n1/pub/75-006-x/2019001/arti cle/00006-eng.htm.

CHAPTER SIX

Home, Hoe, Horse and Hammer? How to Learn from and Live on the Land

ROGER BOSHIER

LIFE ON THE LAND

Planet earth is broken and extreme climate events, global pandemics, racism, terror attacks, and mean-spirited buffoons posing as politicians are among many symptoms of planetary distress. Fortunately for most people, the United Nations is a friend of planet earth and supporter of citizens wanting to learn from the land or ditch urban slums or apartments and move to a piece of rural paradise. The UN also supports Indigenous people who, long ago went looking for bright lights in the city, but now need to find their ancestral land and learn to live there like grandma or grandpa.

In 2015 Murray Sinclair and the Canadian Truth and Reconciliation Commission (TRC) published their nine-volume condemnation of cultural genocide orchestrated by residential schools and there were 94 "calls-to-action." Regarding residential schools and other colonialist horrors, there was a sense of "never again." Regarding the need to respect Indigenous lifeways, Sinclair and his Commissioners said tread lightly. Learn how to live on the land but, once there, learn from the land. This is an important but interconnected distinction. As Taiaike Alfred (2009a; 2009b) and other activists are prone to say "it's always about the land."

At roughly the same time as TRC (2015) reports landed on desks, world leaders were agreeing to take action concerning 17 UN sustainable development goals (the SDGs) (United Nations, 2015). If achieved, the SDGs could reduce carbon emissions and the number of extreme climate events (like fires, tsunami, hurricanes, drought and flooding). In 2020, there was a global pandemic, 4.1 million acres in California were consumed by fire and the UN said "it is now crunch time" and we need everyone to come together.

Goal 15 of the SDGs concerns "life on the land" and complained of deforestation, loss of natural habitats and land degradation. By 2020 Goal 15 was accompanied by "targets" that, if reached, could protect and restore life on the land. The targets were:

- Conserve and restore terrestrial freshwater ecosystems
- End deforestation and restore degraded forests
- End desertification and restore degraded land
- Ensure conservation of mountain ecosystems
- Protect biodiversity and natural habitats
- Promote access to genetic resources and fair sharing of benefits
- Eliminate poaching and trafficking of protected species
- Prevent invasive alien species on land and in water ecosystems
- Integrate ecosystems and biodiversity in governmental planning
- Increase financial resources to conserve and sustainably use ecosystems and biodiversity
- Finance sustainable forest management
- Combat global poaching and trafficking

INDIGENOUS RESURGENCE

For about 150 years, colonizers caused immense social suffering in Canadian Indigenous communities (Irlbacher-Fox, 2009), pilfered land, separated native children from families and incarcerated them in evil schools. Residential school survivors (and their offspring) suffer deep trauma. Now, the Canadian prison system is the new residential school.

Traumatized elders, Indigenous activists and leading scholars have lost faith in the relationship between Indigenous people and settlers (Asch, Borrows & Tully, 2018). Not long ago, the Indian agent, lawyer or Hudson's Bay official needing food, labour or signatures on a contract, figured bribery was best, and brought blankets and beads to "negotiations". The modern equivalent of old-style bribery is a new truck for the Chief, an envelope containing money or share certificates for a fracking operation or an oil or gas pipeline running over Indigenous land. The settler sales pitch is always the same. "In the national interest. Jobs for your people. Revitalize the local economy. Sign this as an act of reconciliation."

Indigenous resurgence stands on two legs. The first idea is to face facts. After 150 years, interacting with settlers has produced dismal results for Indigenous people—poverty, ill-health, depression, high suicide rates, a loss of land. For modern Indigenous people, humiliation and discrimination is an everyday occurrence. In these circumstances, would it be better to tell colonizers to go-to-hell, revitalize Indigenous culture and live on the land like the ancestors? This is the disengagement option.

Disengagement advocates have little or no confidence in the ability of Indigenous people to flourish inside the vast complex of circumstances and actions oppressing their culture and knowledge systems. Many Indigenous people are no longer willing to listen to discredited arguments about the alleged benefits of colonization and "living like us." Moreover, they can no longer be bribed with alcohol, beads, and blankets. Is this the time to shoot-through? Honour the land, regard animals as relatives, gather medicines in the bush, and live in harmony with nature? (see Alfred & Corntassel, 2005; L. Simpson, 2017).

In the second leg of Indigenous resurgence, there are the same disappointments concerning 150 years of humiliation, abuse, and trickery. However, instead of severing relationships with settler institutions and cultures, how about finding fertile nooks and crannies wherein we can work together to build a better world? This is the engagement option.

Put Mohawk activist Taiaieke Alfred (2009a, b, 2015) in a university lecture theatre and, when he says "come on … let's go back to the land," there are willing listeners. However, sitting in the front row are Indigenous youngsters happy to visit distant grandparents but, once out of range of WIFI, city friends and big-stadium rock 'n roll, realize they have no desire to shoot animals, scrape stinking moose hides or suffer the hardships of rural life. Women have additional reasons for reluctance. Rural people have guns which are too often involved in domestic violence. In rural areas, women lack social services and protections found in cities. It's not so easy!

Even fervent advocates of "back-to-the-land" living realise there are not many people on their team. "The bulk of our people have been disconnected from their natural environment on reserves or raised in the city and lack the knowledge and fortitude to live the kind of lives their ancestors did; going "back to the land" is not a compelling call for our people anymore" (Alfred, 2009b, p. 211).

SETTLER URBAN PARADISE LOSES LUSTRE

After World War II, bored (and unemployed) rural dwellers headed to jobs and bright lights in towns or cities. For example, In New Zealand, the post-war Maori urban migration had a profound impact on the meat industry, tribal governance, national politics, and the way life is lived. In China, there are thousands of abandoned rural villages. Chinese teenagers, young parents, and middle-aged farmers recently put away farm tools and headed to factory jobs in coastal cities. Back in the village, grandparents were left to mind babies—many of whom grow up feeling lonely, abandoned, and unloved. The China rural-urban migration is the biggest in human history and still continues.

By 2020, many cities were no longer the paradise once imagined and settler and Indigenous citizens were once again packing bags and looking for rural land with good air, cheaper prices and ancestors buried nearby. In settler societies, privileged people long ago discovered the merits of having a second house at the lake, seaside or up a mountain.

There is also a worldwide tendency to turn rural land into resorts, golf courses, gambling centres, or hideaway retreats such as those in Grand Teton (Wyoming), the richest and most unequal county in the USA where the average income for the top 1% is $28.2 million (Farrell, 2020). Much to the chagrin of the author one of his most loved and respected childhood hangouts near the gannet colony at Cape Kidnappers in Hawke's Bay, New Zealand, is now a "world-class" golf resort with helicopter landing pads and locked gates.

INDIGENOUS URBAN PARADISE LOSES LUSTRE

In Canada the TRC (2015) stood for truth-telling and reconciliation. But there can be no reconciliation without land and, eight years after citizens got their first look at TRC reports, the news is not good. "Truth" is hard to find and "reconciliation" even harder. Instead, there are persistent complaints about "boil water" advisories on reserves, high rates of disease, incarceration, gendered violence, racist police and mediocre school marks. So, instead of enduring urban racism, poverty, and police violence, how about buying food for the horse (or gasoline for the truck) and heading to the ancestral homeland at the head of the fjord or on the other side of the mountain? The fishing is good and there are relatives nearby.

CLIMATE CHANGE AND FOOD INSECURITY

Like the settler en route to the newly bought hobby farm, the Indigenous family heading to their new life on the land, both buy city food to tide them over the first 4 or 5 days. Once this food runs out, what happens next? Neither the Indigenous family or hobby-farmer have a nearby store, and, in both cases, the alleged road is a poorly maintained pothole nightmare. Last summer a bushfire took out wooden bridges, burned the so-called school, and scared away newcomers thinking of opening a store.

Climate change means Indigenous people too easily get into situations where finding food is a problem. When good food is scarce, health problems are not far behind. During the COVID-19 pandemic, people told to stay home in crowded rural houses could not engage in social distancing or do "essential travel" to get food. Climate change meant there was melting ice on hunting routes, expected plants did not appear and there were few (or no) fish at the usual spots (Human Rights Watch, 2020).

Because of global warming and rising waters, hunters fall through ice and homes must be relocated. Canada is warming at twice the rate of the rest of the world and, in the far north, it is warming even faster. If ice roads fail, there are no food deliveries. Ordering food online, curbside pick-up, delivery by drone, or other new-fangled ways of filling an empty stomach are not feasible. So, hunt game, go fishing, and forage for wild berries. Or find cash to buy stale and over-priced produce at a store that imports supplies from southern Canada. When neither of these options are viable, hungry people resort to high calorie junk food which nurtures obesity and aggravates existing conditions like diabetes.

LAND BACK

Whakapapa (genealogy) and land are a fundamental attribute of individual and collective identity in all parts of the world. Hence, Indigenous people have good reasons to learn from the land. Learning to live on the land is a different matter because, in most places, settlers have successfully disconnected Indigenous people from their land.

The main problem is somebody else (such as a foreign oil company or the Crown) claims to "own" unceded Indigenous territory. Unlike the situation in New Zealand, where many Maori chiefs signed the 1840 Treaty of Waitangi, in Canada few Indigenous people signed treaties and their land was not legally given, sold, ceded, or otherwise transferred to others. Instead, the busy Hudson's Bay "officer" or other servants of the British Crown found it easier to bring alcohol and blankets, steal the land, and leave. Where money was used to seal the deal there was not much of it. These were corrupt processes because western notions of individual buying/selling mean little (or nothing) in Indigenous societies where there is collective "ownership" of land.

"Land Back" is a short slogan but captures the goal of more than a century of Indigenous struggle. Architects of the Land Back movement want nothing to do with establishment Indian leaders or organizations prone to surrender land in exchange for one-off payments, share certificates, or a new truck for the chief (Manuel & Klein, 2020).

UN DECLARATION ON THE RIGHTS OF INDIGENOUS PEOPLE

The UNDRIP is a powerful, elegant, well-written declaration concerning land and other rights of Indigenous people. But, if governments (e.g., Ottawa) do not make existing (or future) legislation compatible with UNDRIP, there could be a major loss of traction. While Ottawa prevaricates, in British Columbia, Canada, the John Horgan (New Democratic) government agreed to abide by UNDRIP expectations and demands. For the United Nations, the UNDRIP was a diplomatic triumph and the fact it was crafted by Indigenous people from diverse settings gives it immense credibility (United Nations, 2006)

Canadians have seen dramatic news footage of police dragging Indigenous protesters away from petrocapitalist graders, bulldozers, and other machines carving pipeline routes across unceded Indigenous land. The federal government pays a large political price when uniformed "royal Canadian" police or hired "guards" put handcuffs on Indigenous men, women, and kids and drag them off their land—so oil or gas companies can pollute rivers, mow down burial grounds, destroy fish, and mangle traditional food-foraging places. By 2024 most people knew the UNDRIP forbids resource extraction on Indigenous territory without the written (i.e., legal) consent of landowners. Having a chat, going out for coffee, admiring the river, or sending agents (or lawyers) to "consult" unruly natives does not meet the requirements of the UNDRIP. Even the Prime Minister of Canada needs to realize "consultation" does not denote "consent".

VIEW FROM THE FLOAT PLANE

In Canada the venerable Beaver bush plane conveys freight and people to remote islands, mountain lakes, fjords, bush camps, settlements, cabins, industrial sites, ski hills, and places of refuge. The six-cylinder rotary engine makes a racket and, for the visitor, the best place to sit is next to the pilot.

Look out the window and, in all four directions, there are two clear questions concerning life on the land.

- What can be done with urban-based Indigenous people wanting to reconnect with ancestral lifeways and learn from the land? "Just do it" is a good answer to this question. But, as shown below, it is a complex and multifaceted question.

- What can be done with settlers and Indigenous people wanting to learn to live on the land? "Just do it" is a good answer but, as demonstrated below, there is more to it.

LEARNING SETTINGS

Throughout the 17th, 18th, 19th and most of the 20th century farmers and other rural people introduced family (or clan) members to life on the land at a very young age. Children learned to ride horses, manage animals, hunt game, grow crops, cook, fix defective tools or machines, and build farm structures. Psychologists call this "situated," "experiential", or "social" learning. With grandparents involved it was also an example of very important intergenerational learning.

The best way to learn how to live on the land is to find a suitable place, embrace a lifelong learning ethos, and make the big move. John Dewey (1916) claimed "learning by doing" was the preferred way to learn what matters and what doesn't.

In the learning society advocated by UNESCO (Boshier, 2012, 2018; Faure, 1972), authors expressed grave misgiving about top-down teacher-centred formal education and called for renewed commitments to learning in a broad array of out-of-school settings. Most notably, they envisaged a "learning society" involving active learning in formal, nonformal, and informal settings. In this framework, formality does not refer to pedagogy—the teaching or learning process—it only identifies the place or setting where learning occurs.

The formal, nonformal, and informal trichotomy was created to challenge (or extinguish) the idea all (or most) learning and education occurs in formal settings. If someone could compare the amount of learning occurring in informal, nonformal, or formal settings, as shown in Figure. 6.1, the informal domain would be the clear winner. Learning in formal settings (e.g., schools, colleges, universities) can be important but, as shown in Figure 6.1, compared to what occurs in nonformal and informal settings, there is less of it. The circles represent the size of the footprint occupied by the three learning settings.

The most important things in life are learned in informal settings. In addition, on planet earth there are significantly more nonformal than formal settings. However, the boundaries between these settings are porous and there are numerous examples of formal education outfits that dress themselves to look like they reside in the nonformal domain. For example, university adult education or extension operations are located in formal settings but, with outreach and community organizing, often act and look like a nonformal provider. The boundary between formal and nonformal settings is more porous than the one separating nonformal and informal settings.

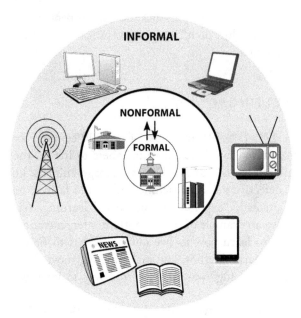

Figure 6.1. Settings for learning from and how to live on the land.
Source: Author

Although the Canadian Truth and Reconciliation Commission (2015) published 94 calls-to-action, their focus was on formal educational settings and, in a stunning omission, lifelong learning, adult education or UNESCO's learning society, learning cities or villages did not make it into their index. Regrettably, several TRC calls-to-action pertaining to learning are so "old school" they could become more of a problem than a solution. Eight years after TRC reports appeared, university statements about these matters are loaded with platitudes, forlorn hopes concerning teacher education and top-down ideas about mandatory courses on Indigeneity. After 2015, universities launched MOOCs (Massive Online Open Courses) on Indigenous topics which promise a lot but deliver little. Not long ago, universities extinguished adult education and other units where nonformal or nonformal learning settings were a research and policy priority and, at the same time, failed to support Tony Bates and other pioneers of online learning.

For many years, Canadians have been at the forefront of UNESCO attempts to nurture out-of-school education and build vibrant learning societies but, with few exceptions (such as the University of Victoria), Canadian Deans of Education are obsessed with teacher education, very poorly informed about distance education or learning out-of-school and, when pressed, resort to feeble clichés like "teacher training" and "Indigenizing the curriculum."

University faculties of education pour vast resources into public relations but do little to nurture "truth" or "reconciliation". In contrast, UNESCO is committed to nurturing learning villages, towns, and cities (see Boshier, 2018) and stands ready to assist Indigenous efforts to infuse a culture of learning into daily life and overcome the challenges of remoteness and living on the land. UNESCO has also stimulated exuberant creativity in China's embrace of learning society thinking and rollout of "learning villages"–such as the wildly innovative Shuang Yu (Boshier, 2012; 2018; Boshier & Huang, 2007) that sits at the interface of Chinese tradition and modernity.

SHAWN AND THE BIG SACK

From 2001 to 2003 Shawn Atleo was a graduate student at the University of British Columbia (UBC) in the Department of Educational Studies–Adult Education. On 23 July 2009, Shawn Atleo ran in an election intended to identify the next Chief of the Canadian Assembly of First Nations. At campaign rallies and all-candidates meetings a large sack of important issues were front and centre—the need for female leadership, derelict and crowded housing, missing and murdered women, boil water advisories, mediocre schools, ill-health, racist police, domestic violence, corruption, petrocapitalism, 1763, dodgy land deals, booze and drugs, disappearing fish, tuberculosis, political malfeasance, and so on.

Shawn did not discount issues brought up by other candidates. However, he decided the most important idea was missing. His response (paraphrased) was like this. "All those issues are important. But there is a solution that pervades them all. What is the one thing all those issues have in common?"

After shrugs and curious looks from opponents, Shawn said,

"they all involve the need for ..... (long pause)."
"LEARNING!"

"LIFELONG LEARNING!"

"I did not say education. More schools and colleges are not the issue."

"I said LEARNING–in all walks of life. Lifelong and lifewide. Learning– everywhere and all the time.

Come on, let's get started."

In 2009 there were eight rounds of voting. But, in the end, UBC (and UTS, Sydney) graduate student Shawn Atleo became Chief of the Canadian Assembly of First Nations.

On 19 July 2012, Shawn sought re-election and new issues demanded attention. However, lifelong learning prevailed again, and, after three rounds of voting, Shawn Atleo again led opponents across the finish line.

Where people live on but do not learn from the land it can have devastating consequences. In Bolivia, impoverished highlanders moving to lower-level grassland or forest know little (or nothing) about slash-and-burn methods of clearing land. Hence, more than 8000 square miles of virgin forest has been lost to fire (Guillermoprieto, 2020).

In 2020, ferocious fires took out 4.1 million acres in California, killed 31 people and destroyed or damaged 10,000 buildings. Every year, Singaporeans, Malaysians, and others are forced to endure thick smoke from Indonesian stubble-clearing fires lit by poorly informed farmers. In Australia, Canada, and the USA there are conflagrations started by selfish, morally flawed citizens with little knowledge concerning life on the land. John Steinbeck ensured moving west would become an enduring theme of California life and culture. Now, fleeing flames and obeying evacuation orders is more important than The Grapes of Wrath.

FORMAL SETTINGS FOR LEARNING

Schools, colleges, polytechnics, technical institutes, universities, and wananga are examples of hierarchical, mostly government controlled formal learning settings. If you are reading this book you probably sat in formal learning settings and have recollections of the best and worst times in classes taught by your favourite or most despised teachers.

In most countries there are opportunities to sit in classes within a formal setting where you might learn from and how to live on the land. However, before handing over entrance fees and sitting through PowerPoint tedium, ask if this institution will help you understand extreme climate events, wildfires, predatory seed companies, and new-fangled cropping practices. Also, does anyone mention unstable (or missing) WIFI signals in rural areas, hospital closures, police brutality, racism, landslides, and absence of museums, art galleries, or libraries on your side of the river?

STEM subjects—science, technology, engineering, and mathematics—all pertain to the ability to learn from and how to live on the land. However, life on the land requires a conceptual apparatus significantly more subtle and larger than the territory encompassed by STEM. Some analysts dance around this problem by adding an "A" for "arts" (so STEAM). But when trainee engineers are told to enroll in mandatory "arts" courses, there will be grumpy resistance. Dammit, I am an engineer, not a poet or a painter!

In China there is an urgent need to have high-party officials and engineers realize dams, railways, irrigation systems, sports arenas, and other giant projects disturb communities and upset life on the land. If in doubt, check out the Three-Gorges dam monstrosity on the upper Yangtze. When the author tried out this idea at Tsinghua university there were smiles from rural students and snorts of derision from trainee engineers. Put an "A" in STEM—you must be kidding? If it's pipelines versus poetry, I'll opt for big machines, fast trains, and tricky welding.

By 2020 there were about 20 Indigenous-controlled colleges, institutes, and universities in Canada. Most were small and vocationally oriented. Even so, resurgence advocates like the idea of stand-alone Indigenous universities which, as demonstrated by Maori wananga in Aotearoa/New Zealand, can be good places to work, attract many students, offer advanced degrees (like the PhD) and celebrate arts and culture.

NON-FORMAL SETTINGS FOR LEARNING

"Non-formal" does not refer to the teaching, learning, or pedagogical processes. Non-formal are out-of-school settings but, in their learning processes, can involve methods, techniques, and devices like those seen in formal settings.

Prisons, churches, political action groups, Indigenous leadership organisations, community centres, clubs, NGOs, libraries, and museums are all nonformal learning settings. Here are examples of nonformal settings where, amongst other things, people can learn from and how to live on the land.

Dechinta Bush University (Canada)

The Dechinta Bush University at Blachford Lake in the Northwest Territories is a nonformal land-based setting. They are not against classroom lectures or discussions and, like other universities, demand written work from students. However, as soon as possible, learners are asked to hike over hills, hunt animals, and net fish in a lake. As day turns to night there is much to learn from the dramatic and magical aurora borealis. Dechinta sits on persuasive bundles of anti-colonial, emancipatory, adult education, lifelong learning, experiential, place-oriented, and situated social theory.

After Y2K worries evaporated, Yellowknife activists pushed for an Indigenous post-secondary institution but ran into resistance from rightwing politicians, petrocapitalists, and neoliberals who felt advanced education might overly excite natives and impede petrocapitalist projects. However, if Dechinta critics thought it would be a simple task to squash the bush university, they under-estimated the social commitment and intellectual acumen of Erin Freeland Ballantyne (2014). She is an Oxford-educated Rhodes Scholar from Canada's north who writes smoothly and raises important questions concerning self-determination and colonialist campaigns to sever Indigenous connections to land.

Dechinta got going in 2010 but, because of financial and other stress, could easily have gone over a cliff (Dechinta Bush University: learning off the land, The Tyee, 22 October, 2013). Moreover, not long after they started, Dechinta faced a moral dilemma when, on 29 April 2011 future king William Windsor married Katherine Middleton. Moose are not where they are supposed to be and now we have to worry about the future king!

Yellowknife is not a renowned honeymoon destination but, in early July 2011, the newly married royal couple landed at the airport. Somewhere in the labyrinth of Buckingham Palace or dreary Ottawa, protocol handlers had decided the royal couple needed to see Dene youngsters celebrating their culture and, at Yellowknife airport, there was a motorcade ready to get the show on the road.

Hearing they had been short-listed for a royal visit, Dechinta board members had to juggle the contradiction of hosting the future king (and his wife) at a place where students and academics scrape moose hides and crank out critical articles about dodgy fur traders and unmet royal promises. Was there anyone in Yellowknife who could reconcile both ends of a complex talking stick? Should us anti-colonists hang out with the soon-to-be king of a fading empire? William is a direct descendant of King George III who, in 1763, made solemn but unkept promises to Indigenous people. However, after a lot of discussion at Dechinta, there was a consensus—yes, host the royal couple and the future king might learn what it is like to be a colonized Indigenous person in Canada.

The big day arrived in the first week of July 2011. After a few functions in Yellowknife, William and Kate got in a floatplane and flew to Blachford Lake (on Great Slave Lake), came ashore and met Indigenous military reservists. They then examined Dechinta student handiwork, inspected moose hides, sat at the outdoor fire-pit discussion place and, for 30 minutes, chatted with local youngsters about colonization, settler-Indigenous relations, and learning from the land.

It was a big deal to have a discussion on Crown–Indigenous relations with the future king sitting on a log. There was no chairperson, no agenda or minutes, but plenty of goodwill and laughter. Dechinta language experts then helped the couple learn greetings in the local language. Elders were arriving, the royal visitors looked like it was their final exam and, grinning sheepishly, said "hello", "pleased to meet you", "thank you", and other things in the Weledeh language. Later, being able to say "please" and "thank you" would influence what they ate for dinner (supper).

After the session at the fire pit, the royal couple spent an hour with Francois Paulette, Dene elder, former chief and Dechinta tutor. Francois, William, and Kate discussed Dene history, Anglo-Indigenous relations, planetary distress, and spiritual significance of Indigenous connections to the earth. Francois then winked at William who knew it was now time to wave goodbye to the Minister of Defence, dismiss bodyguards, get into a canoe and start paddling.

William was in the bow, Kate amidships and Paulette in the stern of the canoe. Everyone wore a Personal Flotation Device (PFD).

"How do I address you?" asked Chief Paulette.

"William and Katherine," said the future king.

"What about you?" asked the royals. "Should we address you as chief?"

"No, call me Francois," said the mellifluous voice in the back of the boat.

Francois Paulette is an eloquent, knowledgeable, and compelling speaker and, being in a canoe with the future king was a chance to continue the earlier discussion. In 1763 King George III (of Britain) recognized Indigenous rights to land natives occupied in the place that became Canada. In addition, George III promised not to "molest" natives "as long as the grass grows, rivers flow, and sun shines." This solemn promise lives on in Article 35 of Canada's Constitution Act of 1982.

Having promised friendship, cooperation, and a "hands off" attitude concerning land and Indigenous lifeways, from 1763 onwards, the honour of the British crown was at stake. However, between 1763 and the 1867 launch of the Canadian "confederation," royal proclamations were tossed overboard and replaced by dodgy dealing, violence, deception, dispossession, and repeated attempts to orchestrate cultural genocide against Indigenous people.

Back in Britain, the subjects of King George III liked him. But, in Canada, Hudson's Bay Company officers and other servants of the British Crown routinely violated royal proclamations and treaties. Modern television pictures of "royal mounted" police assaulting and handcuffing Indigenous people suggest that, since 1763, not much has changed. However, despite disappointments of history and alleged virtues of parliamentary democracy, in 2023 many Indigenous Canadians still believed it is the duty of the Crown to sort out their issues. If there is a problem, ignore your Member of Parliament and call the king or queen

Francois had been a Canadian delegate to early efforts to produce the UNDRIP. Now, having spent an hour chatting on land he was getting into a canoe with the grandson of Elizabeth, the reigning sovereign. William has a Master's degree in Geography from St. Andrews, a good university in Scotland. Kate has a Master's in art history from the same place.

The Prince and Duchess were good paddlers, the canoe moved nicely, and, in the stern, Chief Paulette wondered if the bloke in the bow realized there was a need for contemporary ethical action by the Crown. Could there be another royal proclamation disavowing the doctrine of discovery and agreeing to govern in a manner congruent with the UNDRIP? William ... Kate ... what do you think?

Paulette felt everyone in the canoe understood each other, "a bond was formed" and the royal visitors said they would like to return to the North West Territories in winter (English, 2011). With William's background in geography, Kate's knowledge of art history, and Paulette an expert on northern lifeways, both ends of the toko toko (talking stick) joined together. Out on Blachford Lake, Chief Paulette was learning from well-educated young people and they were listening to (and responding) to him. The Duchess was as fully involved as was Prince William.

At Eagle Island, canoe occupants clambered ashore, looked around and chatted with a chef who had an open fire going and was preparing food. Sunset would not be until midnight and, after a meal of cranberries, bannock bread, caribou, and whitefish there were three hours of "alone time."

This was an historic and mutually enlightening encounter involving Dene and the Crown. Paulette later said Katherine was interested in the historical dimensions of the situation and said she understood spiritual connections to land and Indigenous worries about air, water, animals, and other elements of the natural world. If the future King William tackles only some elements of Paulette's political economy lesson, will canoe-diplomacy nudge Nixon and Chairman Mao's ping-pong exchanges into the back pages of diplomatic history?

Not long after their 2011 trip to Canada, William and Kate had a son named George. How long before he could neatly write numbers–1763–and know what they mean? By 2020, George and his sister Charlotte were at Thomas's Batterseas's school in South London where there was a focus on thinking and questioning and getting kids out of classrooms and into outdoors activities like climbing, sailing, water sports, and first aid. When choosing this school, were his parents thinking about Blachford Lake, the aurora borealis, and sound of paddles working together? (New skills for Prince George and Princess Charlotte, The Guardian, 6 January 2020)

After the royal visit, Dechinta staff, students, and allies focused on options for the future and consistently demonstrated the bush university was fit-for-purpose. With TRC (2015) calls-to-action a compelling backdrop, one year became the next and, in 2019, federal authorities found $13 million for the next five years at Dechinta. Never under-estimate canoe diplomacy!

Critics of the bush university still yapped like barking dogs but, had they read Erin Freeland Ballantyne's (2014) elegant socio-historical analysis, would realize Dechinta is an anti-colonial project designed to foster Indigenous self-determination. Unlike Canada's "big-three" universities (Toronto, McGill, and UBC) Dechinta is not anchored in the prevailing certainties of objectivist ontology and positivist epistemology. Unlike the University of Toronto, Dechinta does not run expensive full-page newspaper advertisements bragging about its achievements concerning STEM. So, for fracking, petrocapitalism, and pipelines, go that way! For Indigenous self-determination and learning from the land, go to Blachford Lake.

Moeawatea Process (New Zealand)

Heritage conservator Dave Harre heard there were plans to demolish the North Island country cottage of famous New Zealand poet Rewi Alley. With government Labour Department help, and wood salvaged from a demolished house in Patea, Dave got building materials organized, arranged vehicles and took 20 at-risk youngsters, tools, food, cooking pots, and sleeping bags into remote, inaccessible, and bush-clad hills behind Waverley, New Zealand.

The cottage was derelict and, even inexperienced youngsters realized bringing it back to life would require commitment, teamwork, and a willingness to work with recycled materials. Among the youngsters were drug-takers, people well-known to police and magistrates, and survivors of insecurity, family tumult, and nasty schools. Arriving at the derelict cottage, there was stunned silence and only one comment.

"Holy shit!"

Then questions started. Where do I sleep? What's for breakfast? Who does the cooking? How do we wash dishes? Did someone bring a tea towel? Is there any soap? Will I be hurt by the cow at the fence? Are there possums around here? Is that the long-drop toilet? Why is that owl (ruru) over there? How do you pull out rusty nails? How long before we see a kiwi? Is there electricity? How long will it take?

On this job-site, learning was always in the foreground and, drawing on his knowledge of Maori tikanga (customs), 3rd force (humanistic) psychology, adult education, lifelong learning, and Rewi Alley's notion of "gung ho" (working together), Dave Harre nurtured youthful workers and, when necessary, shared his extensive knowledge of heritage construction, tools and, ICOMOS (International Covenant on Monuments and Sites). Dave was more of a learning facilitator than a teacher and determined to avoid anything resembling the violence and judgmental authoritarianism of New Zealand schools.

At first, inexperienced youngsters served disgusting porridge and burned toast and the project moved to the edge of a precipice. But Toni Leen had identified with the project and, after a quick consultation with others, built a pop-up country "kitchen", organized shelves, improved water fetch-and-carry systems, and changed rubbish disposal arrangements. Because Toni—later assisted by Hugh McFarlane—served attractive and properly cooked (or not cooked) food, other workers helped in the "kitchen."

For Toni and Hugh, running the bush kitchen in isolated Moeawatea would lay foundations for a massive adventure and career change on the road ahead. How could Toni know cooking porridge and toast for Sandy and Daryl at Moeawatea would lead to making breakfast for Metallica, Shania Twain, or Bob Dylan?

Dave Harre was vegetarian and pleased to have Toni and Hugh producing magical meals and culinary surprises for everybody. At night, well-fed workers listened to foraging kiwi, hooting owls, and sounds coming from deep in the land. Now well fed, along with hammering, checking levels, and pulling nails from old wood, Dave Harre sanded and poked at the theoretical foundations of the project. He knew how to link theory and practice and had Carl Rogers, Abraham Maslow, Edgar Faure, or Paulo Freire struggled up Kohi Road and found Moeawatea they would have heard their concepts in conversations and seen fragments of their theorizing hanging on the house. Time for "conscientization"–where's the coffee?

Most Moeawatea workers hated the formal school system so, in the Waverley back-blocks, "education" was a swear word. Their main preoccupation was "learning" and, during tea breaks and other forms of relaxation, there were animated (and theoretically provocative) discussions about Gung Ho forms of learning, lifelong learning, experiential learning, reciprocal rights, responsibilities nested in Maori whanau, and why it was important to not confuse learning with training or the violent authoritarianism of formal education in New Zealand schools. Youngsters also talked about the need to experience other cultures and get the big O.E. (overseas experience). Had Carl Rogers persuaded a horse to bring him up the greasy hill to Moeawatea, he would have liked what he heard but have a hard time finding a coffee like those in La Jolla.

After 4+ months, the job was finished, the cottage looked magnificent, adjacent land was clean, and a helicopter brought Minister of Labour Phil Goff to a ceremony where, following Indigenous traditions, the new era was signaled by lighting fires. Moeawatea youngsters took what they learned and some established successful careers in New Zealand or abroad. Because of their tenacity, resourcefulness, food-handling skills, and ability to work in threadbare high-stress environments, Toni Leen and Hugh McFarlane moved to Australia and then the U.K., took up catering, and traveled the world cooking for big-time entertainers—such as Metallica, Sting, Shania Twain, Rush, Red Hot Chili Peppers, The Eagles, R.E.M., Lili Allen, Donny and Marie Osmond, Ben Harper, Josh Groban, and Bob Dylan. Who knew "Shania" was really "Eileen" and Sting's name is "Gordon?"

After years on the big-stadium concert circuit in Europe and North America, Toni and Hugh yearned for a piece of land in New Zealand, a supportive whanau (extended family), whitebait fritters and the comfort of cats. Can we go home, wield an axe, grow food, breed hens, collect eggs, and fuss over friendly cows? With lying "leaders" at Westminster and the disunited kingdom staggering from one crisis to another, Toni and Hugh put their rock 'n roll cookware on a ship, stuffed passports in a bag and headed to Heathrow.

Back in New Zealand, they bought rural land on the Kaipara harbour and were soon catching whitebait, milking cows and growing good food. They got married and reached-out to Moeawatea comrades in far-off places—for example, Sandy is now working in Russian and Chinese orphanages, and Daryl is a professional motorbike rider.

Like North West Territories Dechinta bush university graduates, Toni and Hugh McFarlane say Moeawatea was "transformative" and "changed our lives." This was because of their energy, intelligence, and tenacity and Dave Harre's kindness, knowledge of humanistic theory, ability with tools, and desire to embrace everybody—old and young, Maori and Pakeha, male and female.

At age 80 Dave died and was gently (and reluctantly) buried in warm earth at Oratia. But, through his heritage projects, writing and people like Toni and Hugh McFarlane, the Moeawatea legacy lives on (Boshier & Harre, 2009; Harre & Boshier, 1999). When Rewi Alley lived at Moeawatea after World War I the land was rough, and it was almost impossible to graze stock or do much else. Now, at the dawn of the 21st century, it is still rough and the access difficult. But the cottage is a heritage structure and, because of Dave Harre's intervention and the humanistic Moeawatea process, young New Zealanders exceeded their own expectations by doing useful work, listening to, and learning how to live on the land.

Turtle Lodge International Centre (Manitoba, Canada)

This Anishinaabe lodge in Manitoba has a focus on traditional teaching and wellness and is closely linked to the Canadian National Knowledge Keepers Council. They stress links between human and planetary health and worry about fractures separating people from the land. Turtle Lodge is a nonformal learning setting with well-theorized programs concerning Indigenous links between land, culture, politics, climate change, and the need to guard the earth. Turtle Lodge wants to nurture Indigenous sustainable self-determination in a manner congruent with UN Sustainable Development Goals (2015) and theoretical exhortations of Canadian (west-coast) scholars such as Alfred (2009a, b), Borrows (2016), Corntassel (2008; 2012), Ballantyne (2014), and Coulthard (2014).

Standing inside the octagonal Turtle Lodge building in the Saghkeen First Nation, the visitor quickly understands land sits at the centre of most Indigenous formulations concerning individual and collective identity. Look at the land— who is doing what to whom and why? Turtle Lodge echoes the Nuu-chah-nulth notion of Tsawalk (everything is one) which stresses the interconnectedness of all life forms and need for sustainable and respectful stewardship of resources (see Atleo, 2011).

There are good reasons why, by 2020, the "land back" campaign was gaining traction and influential settlers (such as Naomi Klein) had hitched their wagons to the train. Leaders at Turtle Lodge long ago decided settler colonial/capitalist systems are fundamentally opposed to Indigenous resurgence, revitalization, or "land back" efforts.

Turtle Lodge was built in 2002 and, through the leadership of Elder Dave Courchene, has hosted learning events, conferences, and gatherings wherein Manitobans and visitors' study traditional knowledge about first peoples on local land.

Courchene lit the sacred fire at the U.N. Earth Summit in 1992 and shared stages with the Dalai Lama. Among significant tasks at Turtle Lodge is the search for ancestral footprints which help 21st century people learn how to have healthy relationships with each other and the planet.

Care Farms (United Kingdom)

There are several hundred "care farms" in the United Kingdom that, like Moeawatea, bring at-risk people into a rural environment. At the care farm, troubled children, youth, and adults learn about animal and plant husbandry, horticulture, and land management. These places cater to physically and psychologically disabled people of all ages (From tearaway to tractor kids: putting down roots on a care farm, The Guardian, 12 September 2020).

In most agricultural nations there has been an historic tendency for adult farmers to have kids appreciate the land and get out of bed early enough to milk cows, attend poultry, move sheep, calm restless horses, and do numerous other tasks. In places like New Zealand, it was assumed the kid who "showed an interest" would inherit the farm and, in addition, social workers claimed living on the land had therapeutic value.

From time immemorial, just about every New Zealand child worked at a farm or in agriculture-related industries. By spending several seasons on the cooling floor or in the fellmongery at Whakatu freezing works (in Clive, New Zealand) the author managed to stumble through university. Working on (or near) the land was one of the tested ways to overwhelm teenage angst and find a career trajectory. Later, this ameliorative aspect of farm life was called "nature therapy."

In 21st century New Zealand, if soft money appears, care farms are created. But then disappear. Some are iwi (Maori tribe) based; others are run by ex-military personnel and resemble those in the U.K. In New Zealand, there are also community-minded rural people who welcome needy or curious neighbours for big events like whitebait fishing, sheep shearing, egg hatching, calving or horse shows.

By 2020 many kids had fingers and noses in computer games and social media junk food and shuddered at the thought of inheriting (or buying) the family farm and having to do physical work. Moreover, many 21st century youngsters have dysfunctional families, psychosomatic, and affective disorders, addictions and weak (or non-existent) support systems. In too many cases they assume large brown eggs, fresh milk, and most of what they eat comes from a store. Even older children can be surprised when they see an egg coming out of a hen and look skeptical when someone says cows used to be milked by hand. By hand? What about milking machines?

Although care farms are a vital strand in the British social safety net, they rarely attract the attention of researchers. However, the U.K. Care Farming organization makes the following claims:

- Most care farms involve mixed-groups–including people with mental health issues, learning disabilities, mood disorders, addictions, autism, and social withdrawal issues
- Participants get sent to care farms by social service agencies, family members, friends, or community mental health teams
- Participants in care farm activities improve their physique, mental health, and social well-being
- Among psycho-social benefits are improved self-esteem, better moods, increased personal awareness, enhanced confidence, better work habits and respect for animals, land, and machinery
- Participants typically acquire farming, social and basic life skills which apply in farm and other settings

The therapeutic value of being close to nature and working with other people, woodlots, animals, and earth should not be under-estimated. But, just as significant as individual benefits, for a place like Britain, where land has been fought over and its humps, valleys and rivers tamed long ago, the chance for marginalized people to enjoy working farmland comes close to the preoccupations of several UN Sustainable Development Goals.

Variations on the U.K. care farm model occur in the Netherlands where there are a thousand (or more) green care farms for dementia sufferers. In Alberta and Ontario, Canada, there are care farms for vulnerable people—particularly those on the autism spectrum. It is much the same in Japan—where forest bathing is almost a state religion and people adore bullet train rides to uncrowded countryside. Hot Spring bathing is another Japanese obsession—as is the fondness for earth and animals.

Gabriola Island Recycling Organization (Canada)

Gabriola Island is a 57 kim² gulf island in the Strait of Georgia southwest of Vancouver, B.C., Canada. It is about 26 nautical miles (by boat) from Vancouver and a 20-minute ferry ride from Nanaimo, an old coalmining centre on Vancouver Island. It is one of the ancestral homes of the Snuneymux First Nation, has attractive farms and marinas and a population of about 4000.

There is one supermarket, three small restaurants, only one land-based gasoline station, a volunteer fire department and no big-box stores, fast food joints, or foreign-owned designer coffee sellers. Most notably, there are imaginative arts, theatre, and other festivals and a wildly successful Saturday morning farmers market. In August there is a salmon BBQ and fund-raising "concert on the green" (golf course).

There is little (or no) litter on Gabriola where, not long ago, citizens helped vote a Green candidate into the House of Commons in Ottawa. There is a branch of Amnesty International, a horse-riding club and NGOs committed to helping children and adults learn from and how to live on the land. Like other Gulf Islands, Gabriola is under the jurisdiction of the Islands Trust whose main mission is "to preserve and protect" the natural world.

Roughly 30,000 to 11,000 years ago, Gabriola was covered in thick ice. The first European to arrive was Jose Narvaez, a Spanish naval officer who, in 1791, sailed and rowed through Gabriola's Silva Bay and produced valuable drawings and charts of the area. Captain Vancouver arrived in 1792 and had the delicate task of telling the British Admiralty their navy had been beaten by Spain.

There is no rubbish dump (or landfill) on Gabriola. Stimulated by the UN Sustainable Development Goals and Green Party supporters in Canadian Gulf Islands, in 1989 activists decided Gabriolans needed to "reduce", "re-use", and "recycle" and formed an organization with almost no paid staff but much space for willing volunteers. The recycling yard is at the end of "Tin Can Alley."

By 2020 GIRO (Gabriola Island Recycling Organization) was highly regarded and even affluent citizens avoided expensive off-island department stores and bought clothes, CDs, kitchenware, books, tools, garden equipment, outdoor furniture, bikes, boating gear, and other stuff at GIRO. Everyone loves a bargain and GIRO quickly became a place to recycle old batteries, dirty oil, comatose household appliances, tired woodstoves, and not-so-valuable skis kids left behind when they left home. With treasures like grandma's bone china or cast-iron cookware, GIRO became a tourist destination. Before fishing for salmon, you have to see GIRO!

At the annual Island Fall Fair, couples bring recent clothes purchases for the "GIRO fashion parade." Even staunch island introverts—police, teachers, the post-mistress—adore the drama and theatrics of the GIRO fashion parade.

Although only a small island—dependent on float planes and ferries—Gabriolans are very well informed about international news and assiduously follow UN developments—particularly those involving climate. In 2019, activist instincts at GIRO were on display when the Board decided to launch a "climate change education and action program." Learning from and how to live on the land became a recurring theme in GIRO's climate program.

- Staff are trained to ensure islanders understand what happens to plastic, oil, appliances, building material offcuts, beach debris, clothing, tools, bottles, bikes, outdoor furniture, cardboard boxes, suitcases, and other stuff left at GIRO
- Second, staff write articles for the local newspaper and talk at environment-oriented community meetings
- Third, staff distribute literature about land stewardship and, during debates about proposed "developments," build support for UN SDGs and the GIRO point of view
- Fourth, GIRO was approached by solar power advocates and installed a huge array, inverters, controllers and other gear needed for an optimal "off-the-grid" electrical power plant

Because the author, his computer, radios and coffee grinder depend on solar power, he keeps a close eye GIRO's work concerning alternative energy. Moreover, when very heavy deep-cycle batteries misbehave, they get a truck ride to GIRO and don't come back. As well as being a recycling outfit, GIRO is well on the way to becoming an Adult Learning and Education Centre.

INFORMAL SETTINGS FOR LEARNING

Informal learning settings are serendipitous, casual, and often unplanned. Notice a roadside billboard and the driver learns there is a fire-ban, program change at the film festival, a lost dog, an outboard motor for sale, and change-of-date for a dance at the community hall. Yikes, the dance is now on Friday, not Saturday. Learning in informal settings is often unplanned, sporadic, and inclined to miss (or ignore) important components. Television, newspapers, coffee shop conversations, and the Internet all involve learning in an informal setting.

A good example of using an informal setting for learning started in Tofino (on the west coast of Vancouver Island) but soon spread to other places. Tofino was once a marginal logging and fishing village but developed a hippy back-to-the-earth feel and turned into a favourite destination for surfers, whale watchers, or visitors wanting to experience Indigenous culture.

The Tofino "chicken lady" likes to discuss life on the land and *all-things-chicken*. At her place, she has her own hens and 15 or 20 belonging to neighbours. During COVID lockdowns, people said they missed the chance to talk about chickens. Her response was to stage "chicken socials" over tea and biscuits at the Tofino botanical gardens. Bring a chair, stay 2 m apart, and yell out your questions. How do I clip chicken wings? How do I build a coop? What does a fertilized egg look like? What is the role of roosters? Is chicken manure good for gardens? What is the difference between a Rhode Island Red and a Silkie? This is Snowflake. Is Snowflake a boy or a girl?

There was excitement when two participants fussed over incubated eggs they bought from old Mr. Jones who said "don't worry, any time soon!" Lauren and Colin had incubated eggs in Ucluelet and unwisely counted them before they hatched. When excitement subsided, they had 8 baby chickens when there was supposed to be 12. Oh dear, where is the chicken lady of Tofino? (Renwick, 2021).

Because of their isolation and colonial experience, New Zealand has a large number of self-educated farm-gate intellectuals learning in informal settings. My marriage is crumbling, the horse refuses to move, a grandson was thrown out of school, and tomatoes are being killed by blight. So, go to the back fence and call the farm-gate intellectual—also known as the clever neighbour.

New Zealand farm-gate intellectuals have revolutionized film-making (Peter Jackson), yacht design (Bruce Farr, Laurie Davidson, Ron Holland), America's Cup campaigning (Peter Blake, Grant Dalton), motorcycle racing (Burt Munro, Maury Dunn, John Brittain), food processing (James Wattie), marine jet drives (Bill Hamilton), mountaineering (Ed. Hillary, George Lowe), aerial mapping (Pete van Asch), and numerous other technologically intense, complex, and difficult endeavours. Ed Hillary hated Auckland Grammar and none of these people have flash university credentials. Knowing this, Boshier (2002) asked some of them how they learned what they needed to know. They mostly depended on informal settings. When asked about formal education, yacht designer Laurie Davidson snorted and said "school ... you must be kidding! Just go and talk to your mates! It's a matter of learning by doing."

Peter Jackson did not create King Kong or a truckload of hobbits by studying in formal or even nonformal settings. He has never been to university or a film school. In the early splatter film-making days his only "school" was the basement of the family house in Pukerua Bay. But he understands networking in informal settings. Hence, the night Peter picked-up 13 Oscars for Lord of the Rings was good for New Zealand film but even better for highlighting the power and importance of learning in informal settings.

Learning in informal settings can be fun and addictive. It is always exciting to learn something new and there are plenty of places and opportunities to learn from or how to live on the land:

- Forest and Bird societies
- Hiking, mountaineering, ski and ski-touring clubs
- Forest bathing
- Rural/agricultural fairs and festivals
- Farm clubs (like 4-H in Canada and Young Farmers Clubs in New Zealand)
- Sheep shearing contests
- Farming "field days"
- Logging contests
- Ploughing festivals and contests
- Gymkhana and other horse shows
- Professional or volunteer emergency-response events
- Rural fishing societies
- Country Women's Institutes (Canada)
- Mother's Union (New Zealand)
- Radio, television, and the Internet
- Community billboards and awareness campaigns
- Political demonstrations and protests
- Newspapers and magazines
- Photography
- Backpacking and traveling

Another extraordinary story about learning from the land comes from the storm-tossed west coast of Ireland. Connemara sits in an old, rugged, and historically significant landscape. Tim Robinson was an English cartographer, mathematician, visual artist, and writer. He was fed-up with the cost of living in London, so in 1972 set out to learn from the land in West Ireland, to make maps and write books. Tim and his wife Mairead settled in Inishmore but later moved to a derelict lace maker's workshop on the quay at Roundstone in Connemara. Soon thereafter, Tim, concerned with the broken planet, continued to learn from interlocking inlets, islands, lakes, mountains, peat bogs, cemeteries, and neighbours—dead or alive. When pressed to identify his methodology, he said "long, solitary walks." Robinson pioneered "deep mapping" which starts at the top but dives into geology, archaeology, botany, and the imaginative world of faith and folklore. Tim learned from the local land and, with his wife working alongside, successfully mapped Burren, Connemara, and the Aran Islands. Friends said Robinson heard speech where others imagined silence. A rational visionary, he had the gifts to gather his perceptions in looping arcs of prose, each word a step towards the next, all in synchrony with the life of sea and stone, the dolphin in the cave, the rock-strewn beach. He had determination too, and an attitude towards life that remained curious about the world around him, even to the end (Allen, 2020).

Between 1845 and 1852 about a million Irish died of starvation and another million fled on "famine ships." There were famine graves near Robinson's Roundstone hideout and, on long walks, he sometimes felt like a priest bending his ear to the mouth of a dying man. Along with exquisitely detailed maps, he published acclaimed books on the Arran Islands. Later, *Listening to the Wind: the Connemara Trilogy*, Parts 1–3 (2006, 2008, 2011) won the Irish Book Award for nonfiction and alerted a global audience to the joys of studying local land by listening, finding the right words, and making art. During long walks, Robinson resurrected mostly unknown details concerning ruined cottages and castles, famine graves, wet bogland, rock walls, pointed peaks, the agonies of the past and future threats posed by climate convulsions.

Tim Robinson celebrated the local by listening to the earth and noting its rhythms. But it was not all good news. He picked-up Parkinson's disease, moved to London in 2015 and, on the 3rd of April 2020, died from COVID-19 at 85 years of age. In a 3rd April 2020 obituary in the Irish Times, the headline writer labeled him "the English writer who-went-native in Connemara." His work is a spirited, detailed, immaculate, and breath-taking demonstration of learning from the land by going on a long walk. How long before the amazing Robinson of Roundstone shows-up in storywork at "Musqueam 101" on the Fraser River in Vancouver, in land-focused courses at Awanuiarangi Wananga in New Zealand or in discussions around the Dechinta Bush University fire pit?

REFERENCES

Alfred, T. (2009a). *Peace, power, and righteousness: An indigenous manifesto*. Oxford: Oxford University Press.

Alfred, T. (2009b). *Wasase: Indigenous pathways of action and freedom*. Toronto: University of Toronto Press.

Alfred, T. (2015, November 12). A Talk by Taiaiake Alfred: Research as indigenous resurgence. Lecture to a class at *Carleton University*. Ottawa, ON. Retrieved from: https://www.youtube.com/watch?v=myIUkzbiG_o

Alfred, T., & Corntassel, J. (2005). Being indigenous: Resurgence against contemporary colonialism. *Government and Opposition, 40*(4), 597–614.

Allen, N. (2020, October 26). Saving Tim Robinson's Roundstone home, his last gift to the Irish nation. *Irish Times*, 06.37 sec.

Asch, M., Borrows, J., & Tully, J. (2018). (Eds.). *Resurgence and reconciliation: Indigenous-settler relations and earth teachings*. Toronto: University of Toronto Press.

Atleo, R. (Umeek). (2011). *Principles of Tsawalk: An indigenous approach to global crisis*. Vancouver, BC: University of British Columbia Press.

Ballantyne, E. F. (2014). Dechinta Bush University: Mobilizing a knowledge economy of reciprocity, resurgence and decolonization. *Declonization: Indigeneity, Education and Society, 3*(3), 67–85.

Borrows, J. (2016). *Indigenous legal traditions in Canada: Report for the Law Commission of Canada.* Ontario: Law Commission of Canada.

Boshier, R. W. (2002). Farm-gate intellectuals, excellence and the university problem in Aotearoa/New Zealand. *Studies in Continuing Education, 24*(1), 5–24.

Boshier, R. W. (2012). Lifelong learning as a flag of convenience. In D. N. Aspin, J. Chapman, K. Evans, & R. Bagnall (Eds.), *Second international handbook of lifelong learning, Part 4* (pp. 701–720). Dordrecht: Springer.

Boshier, R. W. (2018). Learning cities: Fake news or the real deal? *International Journal of Lifelong Education, 37*(4), 419–434. doi: 10.1080/02601370.2018.149190

Boshier, R. W., & Harre, D. (2009). A critical application of community service learning in rural New Zealand. In G. Strohschen (Ed.), *Handbook of blended shore education: Adult program development and delivery* (pp. 139–156). Dordrecht: Springer.

Boshier, R. W., & Huang, Y. (2007). Vertical and horizontal dimensions of China's Shuang Yu learning village. *Studies in Continuing Education, 29*(1), 51–70.

Corntassel, J. (2008). Toward sustainable self-determination; Rethink the contemporary rights discourse. *Alternatives: Global, Local, Political, 33*, 105–132.

Corntassel, J. (2012). Reenvisioning resurgence; Indigenous pathways to decolonization and sustainable self-determination. *Decolonization: Indigeneity, Education and Society, 1*, 86–101

Coulthard, G. (2014). *Redskin, white masks: Rejecting the colonial politics of recognition.* Minneapolis: University of Minnesota Press.

Dewey, J. (1916) *Democracy and education.* New York: Dover Publications.

English, R. (2011). Will and Kate witness midnight sunset in northern Canada. *Daily Mail*, 6 July.

Farrell, J. (2020). *Billionaire wilderness: The ultra-wealthy and the remaking of the American west.* Princeton: Princeton University Press.

Faure, E. (1972). *Learning to be.* Paris: UNESCO.

Guillermoprieto, A. (2020). Bolivia's tarnished savior. *New York Review of Books*, 3 December, 56–58.

Harre, D., & Boshier, R. W. (1999). The Moeawatea process: Service learning with kiwi attitude. *New Zealand Journal of Adult Learning, 27*(2), 1–27.

Human Rights Watch. (2020). My fear is losing everything: The climate crisis and First Nations right to food in Canada. New York: *Human Rights Watch*, p. 76.

Irlbacher-Fox, S. (2009). *Finding Dahshaa: Self-Government, social suffering and aboriginal policy in Canada.* Vancouver: UBC Press.

Manuel, K., & Klein, N. (2020). Haggle no more; the push to reclaim Indigenous territory. *The Globe & Mail*, 21 November, p. O1.

Renwick, M. (2021). Chickens come home to roost in a good way during the pandemic. *The Globe & Mail*, 2 January, p. A5.

Robinson, T. (2006). *Listening to the wind: The Connemara Trilogy, Part 1.* London: Penguin.

Robinson, T. (2008). *The last pool of darkness: The Connemara Trilogy, Part 2.* London: Penguin.

Robinson, T. (2011). *A little Gaelic community: the Connemara trilogy, Part 3.* London: Penguin.

Simpson, L. (2017). *As we have always done: Indigenous freedom through radical resistance.* Minneapolis: University of Minnesota Press.

Truth and Reconciliation Commission of Canada. (2015). *The* final report of the Truth and Reconciliation Commission of Canada, Vol. 1 to 6. Montreal: McGill-Queen's University Press.

United Nations General Assembly. (2006). *United Nations Declaration on the Rights of Indigenous People*. New York: United Nations.

United Nations. (2015). *The 2030 agenda for Sustainable Development*. New York: United Nations.

CHAPTER SEVEN

Fostering Growth Through Indigenous and Land-Based STEM Education: Lessons from Plant Relatives

KELLY KING, MADISON LAURIN, & KRISTIN MUSKRATT

INTRODUCTION

It's a warm July morning and TRACKS staff and students are filing onto a school bus to make our way through Michi Saagiig Nishnaabeg Territory, north on Highway 28 towards Woodview, Ontario. We're heading to Petroglyphs Provincial Park, a very special place to this territory, which includes the largest known collection of ancient petroglyphs in Ontario. Also known as Kinomaagewapkong (the Teaching Rocks), this sacred space is under the care of Curve Lake First Nation and is known as a place to learn about local knowledge and ancient understandings of how to live with the land (Williams, 2018).

This is a day we've been looking forward to for months as it is a day where we can truly exemplify what TRACKS is all about. We're midway through our week-long camp and the kids are excited to get away from our regular spaces and visit a new place some of them have never even heard about. In addition to our group of campers, the Oshkwazin Youth Leadership staff are also joining the trip, as well as our Outreach & Education staff instructors. We're equipped with backpacks loaded with lunches, water coolers, extra snacks, and other materials needed for a day shared on the land.

The bus ride is bumpy and loud as the kids play games and chat with their newfound friends. Even though some of these campers have been to TRACKS many times before, they are still excited to see what the day will bring. By the time we make it to the gates of the park, we're all buzzing with excitement, in anticipation of what might come next.

As we make our way down the winding road, the staff encourage the kids to look out into the forest to see if they can spot any animals. The land that surrounds this sacred site is famous for being home to many deer and other beings so it's paramount that we keep our eyes peeled!

We get off the bus and make our way to the interpretive centre. After spending some time learning about the history of this place, we meet up with Beedahbin Peltier to take our learning to the forest. Beedahbin is a medicinal plant knowledge holder and teacher of Anishinaabemowin from Wiikwemkoong First Nation. He also once worked for TRACKS as an instructor and is well versed in teaching kids about how knowledge systems can work together. It's a delight to everyone to have Beedahbin join us for the day to share some of what he knows about local plants, as well as some of the significance of this place to the Nishnaabeg. We gather as a group to greet him and begin to walk down the Nanabush Trail. With an extensive knowledge of native plant species and their medicinal properties, Beedahbin guides us along the trail, pointing out one plant at a time. He even invites us to taste a few and teaches us about which parts of our bodies they can support and how these plants can be honoured and used in our everyday lives.

While we're hiking, Beedahbin shares with us that if we look closely at the land in protected areas like this one, you can sometimes see concentrations of diverse medicinal plant species. This is because of the presence of a long and honoured knowledge system that, above all else, teaches us to live in reciprocity with the land that offers us so many gifts. Holding a flower head of wild bergamot, Beedahbin explains to us that when harvesting medicines, it's seen as disrespectful to hold a flower or seed head upright. If we hold them upside down, we show great respect by supporting the plants in their ultimate goal to drop and spread their seeds. When someone bends down to harvest a different plant, these seeds will be dispersed along the pathways that person makes. Beedahbin explains to us that we can still see traces of traditional homes if we look closely at the land, as the practice of spreading seeds has been observed by the Nishnaabeg for thousands of years. These densely populated areas of medicines are where homes would have stood, their walls lined with plants hanging upside down to dry so they could be used for many health needs and to treat ailments. These concentrations of plants show us how systems of reciprocity have been literally woven into the everyday lives of Indigenous peoples of this territory for time immemorial.

After learning about several more plants along the trail, we head back to the interpretive centre to enjoy cups of wiikehn tea prepared by Beedahbin. Also known as sweet flag, this plant can be found throughout much of this territory and grows in abundance in marshy areas. While some of us enjoy the tea and continue learning with Beedahbin, some of the instructors begin a game of Camouflage in a nearby patch of trees. With the Oshkwazin Youth Leadership staff and the Outreach & Education staff splitting up to facilitate what's needed, it's clear that everyone appreciates being given the choice of what to do next. Even though this is a time for everyone to unwind, we are able to nurture each camper in the way they need to be supported.

Our time with Beedahbin and our plant relatives allows us to begin illustrating the idea of Anishinaabe aadiziwin. This is the embodied understanding that species, elements, and generations are always working in collaboration with each other and are constantly transferring both seen and unseen energy back and forth. Anishinaabe aadiziwin also shows us that responsibility, reciprocity, and affirmations of life are keystones when it comes to Indigenous sciences and understandings of the world (Peltier, 2020).

As an educational youth program, TRACKS has always been a place where seeds are planted. Whether it be the spread seeds of native species or lessons shared by knowledge holders, the kind of programming that is exemplified above opens up new possibilities for nourishment, ideas, and growth for everyone involved. Our day at Kinomaagewapkong teaches us that there are many ways of learning and that if we stop and listen for a while, the land will offer us incredible in-depth knowledge. The land exemplifies to us that growth happens best in the context of deep respect and reciprocity, in the realm of traditional ecological knowledge, and in the footsteps of the ones that cared for the land before us. It's our responsibility to make sure there is foundational support for those seeds and ongoing care and attention for what they need to grow.

CULTIVATING SOILS: WHAT IS TRACKS?

By Madison Laurin, Operations Coordinator
TRACKS (TRent Aboriginal Cultural Knowledge and Sciences) is a youth program based at Trent University in Nogojiwanong (Peterborough, Ontario). Trent University is home to a great deal of innovative teaching and learning taking place in diverse programs and disciplines, including the Indigenous Environmental Studies and Sciences (IESS) program. IESS is a multidisciplinary, academic undergraduate program developed in 1999 by Haudenosaunee scholar Dr. Dan Roronhiakewen Longboat (He Clears the Sky).

Since its inception, a primary goal of the IESS program has been to promote a deeper understanding and appreciation for Indigenous environmental knowledges, and to show the importance of incorporating both Indigenous and Eurocentric scientific ways of knowing into environmental science education. The IESS program takes inspiration from the Two-Eyed Seeing approach to education, a philosophy most often attributed to respected Mi'kmaq Elders Albert and Murdena Marshall. Inherent to this approach, as described by Hatcher et al. in "Two-Eyed Seeing in the Classroom Environment: Concepts, Approaches, and Challenges", is a "respect for different worldviews and a quest to outline a common ground while remaining cognizant and respectful of the differences"

(Hatcher et al., 2009, p. 152). By inviting both Indigenous and non-Indigenous students and teachers to learn together, IESS promotes a deep appreciation for multiple ways of coming to know nature and sets the standard that sharing knowledge is the most effective way to work through environmental challenges facing humanity today. Dr. Longboat has described the sharing of knowledge stemming from multiple ways of knowing as the act of adding varied tools to each of our "tool belts," or personal arsenals of knowledge, that we will rely on to address environmental issues.

As IESS grew as a program, many of the adult youth enrolled in undergraduate programs were becoming increasingly adept at making use of their tool belts, sharing with one another and relying on various knowledge systems to understand the local environment. However, we were concerned around the disproportionately high number of Indigenous youth who were not reaching the undergraduate level or engaging with the sciences along their learning journey. This trend is evidenced by low attendance rates for Indigenous students in secondary school, leading to low graduation rates for these students, and eventually low rates of post-secondary enrolment and certification (Indigenous and Northern Affairs Canada, 2011). Disengagement at the secondary level parallels the low percentage of Indigenous peoples in STEM fields at post-secondary institutions and in STEM-based careers (Statistics Canada, 2016). It became increasingly apparent at Trent University that many Indigenous youth were disengaging with subjects like math and science well before they reached post-secondary education where they would have access to programs like IESS. Furthermore, both Indigenous and non-Indigenous students who enrolled in IESS were entering the program with fixed ideas that certain knowledges or tools, primarily those based in Western sciences, would be most effective in addressing environmental challenges.

In 2010, just over 10 years after IESS was launched, Dr. Dan Longboat, and Dr. Chris Furgal, also of the IESS program by that time, brought together a group of Indigenous and non-Indigenous colleagues to develop the TRACKS Youth Program as a way of introducing youth to Indigenous and environmental ways of learning at a much earlier age. Dr. Longboat is celebrated for his traditional Rotinonshón:ni knowledge, which he embeds into his teaching and in developing the IESS and TRACKS programs. Dr. Furgal is a non-Indigenous, multidisciplinary environmental health researcher and educator who has been working across knowledge systems in partnership with Indigenous knowledge holders and practitioners in support of Indigenous communities, organizations, and governments for over 25 years.

The hope was that TRACKS would create the opportunity for Indigenous and non-Indigenous youth to be involved in a continuous pathway of STEM education, focused on seeing the benefits and potentials of complementary approaches guided by Indigenous and Eurocentric knowledges, from kindergarten through to the post-secondary level. In so doing, TRACKS would encourage youth from diverse cultural backgrounds to be open to a scientific and environmental approach of braiding multiple knowledge systems, and to add diverse knowledge and skillsets on their toolbelts. With this hope of engaging youth at an early age, TRACKS began as an idea for a model of youth education led by youth mentors from within the University, rooted in the need to provide on-the-land learning opportunities, and to highlight these different ways of knowing. Further, the learning and teaching model for TRACKS program relied on cultivating relationships with families, communities, and knowledge holders, including Elders, scientists, and environmental practitioners. In 2012, TRACKS began running camp programs after 2 years of refining content, prioritizing research, and deepening relationships with nearby First Nation partners.

Since then, TRACKS has grown to reach over 5,000 youth per year through workshops, outreach events, and camps. The TRACKS team has also grown from a committed group of volunteer faculty, community members and a small complement of part-time staff, to now employ three full-time coordinators and up to 12 part-time seasonal instructors. Throughout all of this impressive growth, the program has never wavered from its original vision. 2020 marks the tenth anniversary of TRACKS and so much of the program's success today is due to the dedication of those individuals first involved in cultivating the strong foundation upon which TRACKS has been able to grow. The TRACKS model endeavoured to bring together distinct knowledge systems, foster knowledge sharing, and encourage cooperative environmental leadership as central tenets of the program. In addition, TRACKS embraces values such as mentorship and land-based learning—values inherent to Trent University's IESS program. Paralleling IESS, we take inspiration from traditional methods of teaching from many Indigenous communities, which centre mentorship, intergenerational sharing, the nurturing of individual strengths and abilities, and providing rich first-hand experiences on the land. We foster spaces of teaching and learning, where youth are encouraged to seek advice, guidance, and direction from positive examples around them, including the land, older youth, TRACKS staff, Elders, and knowledge holders. As Dan Longboat has said, "IESS and TRACKS provide an outlet for students to be immersed in other ways of knowing and thinking, where the authority of Indigenous knowledge is valued and, coupled with science and technology, leads to innovative problem solving for environmental issues facing our world" (Furgal, McTavish, & Smith, 2018, p. 19). By immersing students in other ways of knowing and thinking, the IESS and TRACKS pedagogy aligns with Two-Eyed Seeing and proves diametrically opposed to that of colonial frontier logics in education, as described by Dwayne Donald in *Decolonizing Philosophies of Education* (2012). In his chapter "Forts, Colonial Frontier Logics, and Aboriginal-Canadian Relations", Donald describes typical educational settings in the colonial context of Canada as perpetuating "naturalized separation" of Indigenous and non-Indigenous peoples and ways of knowing, based on the "assumption of stark, and ultimately irreconcilable, differences" (2012, p. 91) stemming from the "colonial project of dividing the world according to racial and cultural categorizations" (2012, p. 92). TRACKS, aligned with a Two-Eyed Seeing approach, provides an educational setting where Indigenous and non-Indigenous peoples share space and learn from one another, in a way that emphasizes the value of diverse cultures and knowledge systems. The goal of TRACKS is to provide a foundation of learning which encourages innovative ideas to grow, including but not limited to those addressing environmental issues, and it is clear in practice that the best ideas stem from collaboration and sharing across worldviews. As we continue to nurture these foundations stemming from relationships with communities, youth, and land, we work to actively encourage new growth. This growth looks like: building new relationships, reaching more youth, and encouraging more connections to be made between Indigenous and Western sciences.

In the next two sections of this chapter, we introduce the ways in which we as a team of educators encourage growth of learning and leadership of youth through TRACKS programs. TRACKS offer two programs: Outreach & Education, and Oshkwazin. The Outreach & Education program delivers land-based workshops in schools and camp programs in the Peterborough area for kindergarten to grade 12 students. Our camp programs are geared towards learning from the land through local Indigenous knowledges and Western sciences. TRACKS camp invites students between the ages of 6–12. In multi-day camps, our activities focus on offering opportunities for students to be introduced to and to develop intercultural understanding of diverse knowledge systems while being in nature. Meanwhile, Oshkwazin, (an Anishinaabemowin term, the meaning of which and connection to TRACKS will be explained later on in the chapter), is a youth leadership program dedicated for Indigenous high school-aged students. Through Oshkwazin, we encourage Indigenous youth to grow their leadership skills and cultural knowledge through community-based approaches, including lunch socials, gathering events, and a youth employment program. In the next section, we further describe our approaches in these programs in detail.

PLANTING SEEDS: OUTREACH & EDUCATION PROGRAM

By Kelly King, Outreach & Education Coordinator
TRACKS Outreach & Education focuses on engaging Indigenous and non-Indigenous youth in land-based ways of learning from both Indigenous and Western sciences. During our camp days, we often gather in the tipi that is cared for by the First Peoples House of Learning at Trent University. On each Monday

morning of camp, the kids begin to arrive, nervously clinging to their parents who encourage them to chat with other campers and get to know this space where we'll be spending a lot of our week. The instructors welcome the campers with a smudge, explain the protocols of the place, speak about the history of the territory, and play warm-up games to "break the ice". Because the notion of relationships and building a community is important for our land-based approach, we begin our program by developing an understanding of each other and our place of learning.

Within the Outreach & Education program, we're consistently working on how to cultivate and exemplify forms of allyship between knowledge systems and between people. As we work with youth from diverse backgrounds, we feel it is important to mirror and model ways to be a good ally, being respectful towards the original inhabitants of a local space throughout our mentorship practices. Our staff team is composed of undergraduate students, mostly from the Indigenous Environmental Studies and Sciences (IESS) program, who have been learning how to braid multiple knowledge systems in their classrooms and with their peers. Our staff members are both Indigenous and non-Indigenous people, and they exemplify what it means to work towards weaving worldviews and scientific understandings of the world and our place within it. Learning happens from a place of growth and inspiration, where staff—and subsequently participating youth—have their particular skills, needs and interests nurtured because they are surrounded by a community willing to share their knowledges and experiences.

Our program strives to make Indigenous cultural awareness a standard among the youth we engage. We do this by centring the knowledge system of the land on which we work and teach, Michi Saagiig Nishinaabeg territory, and highlighting other Indigenous knowledge systems from across Turtle Island as well. We aim to not only showcase Indigenous understandings of the world, but to honour them as key worldviews for moving forward with responses to the issues we are seeing more of everyday. In many educational spaces, kids are learning about the overwhelming challenges facing our world today including climate change, environmental degradation, the spread of viruses, loss of biodiversity around the globe, as well as polluted waters and lands, to name a few. With readily available information from the media and elsewhere, we are witnessing higher rates of climate anxiety among the youth with whom we work. We believe that if solutions to these problems are not offered as a core part of programming and curriculum, the information on these global challenges can become paralyzing for students. It is our goal to provide education programs that can also act as an antidote to these anxieties by continually centring positive responses based on land-based approaches to problem solving. By connecting youth—and ourselves as educators—to the land, we can find that a lot of these solutions lie in the knowledge of this territory and the experiences and epistemologies of the Michi Saagiig Nishinaabeg. By providing hands-on examples of intercultural understandings within science education, we can structure our programs from a place of regeneration and hope, as opposed to a place of fear.

First and foremost, the land is our teacher. Everything we do in our program is based on territorial understandings of the seasons and lunar cycles, and the corresponding lessons emerging from our more-than-human relatives. Whenever possible, our programs take place on the land—in traditional spaces and shelters, cedar groves, wetlands, and meadows. Beyond being physically on the land, land-based learning also refers to a curriculum that is inspired by what is observed on the land, which teaches reciprocity between humans and our more-than-human relatives. We aim to show TRACKS youth that we can learn just as much from the land as we can from each other. Therefore, our curriculum is developed, refined, and delivered by the staff we have on our team with the understanding that programming may shift and evolve depending on what is happening on the land during any given week or season. This also gives us the opportunity to value and incorporate the knowledge campers bring with them. This means that our programming is flexible and aims to connect the knowledges and lived experiences that all participants bring to these spaces.

As a team, we often have conversations to reinforce our main goal "to plant seeds." Over five days of camp programming, we touch on many topics and interests. It is our hope that the youth we work with will continue on their learning journeys, holding onto one of those seeds; maybe they learned how to successfully start a friction fire, maybe it was a teaching shared with them by a guest Elder, or maybe it was learning how to identify and track deer markings we found on a hike. Whatever that seed of knowledge is, the most important part of this process is to cultivate a learning environment that nourishes those seeds. We know seeds cannot thrive in an environment that does not feed them, so how do we create spaces that provide the nutrients necessary for optimal growth?

By delivering programming that honours multiple ways of knowing, we're also invited to think of the ways in which we *come to know*. This broadens our scope of programming to move beyond the content of our curriculum and encourages a focus on how we facilitate the context of our camps. By fostering an inclusive pedagogy within our program (i.e., honouring youth voices and lived experiences, engaging with multiple knowledge systems, highlighting intergenerational learning), we can work towards providing an experience that allows campers to find the tools they need to build positive relationships with everything around them. Learning in a space that not only prioritizes the braiding of multiple knowledge systems, but also the ways in which we engage with each other and the land, means we can make new connections we can't necessarily find in Western systems which focus on Enlightenment-based educational settings. As Donald (2012) suggested, colonial frontier logics of environmental education place the child as separate from the environment in a way that perpetuates binaries. If we are going to learn from worldviews that resist the notion that we are removed from the environment, we must engage in pedagogies that work outside hierarchical, colonial perspectives (Nxumalo, 2019). We aim to structure our programming in a way that invites each participant to bring their full selves to it and honours campers' life experiences as teachings unto themselves. Our pedagogy includes small group sizes, honouring specific needs and learning styles, learning within culturally relevant and decolonial spaces, providing ample mentorship opportunities, and valuing everyone's experiences as valuable lessons for all. This allows the interests of each child to be recognized and their skills and passions honoured during the week of camp. Our pedagogical approach also recognizes that we do not exist as separate from the land; everything we do has an impact, whether positive or negative, and this needs to be taken into account when it comes to teaching youth in place.

When considering the ways in which our campers learn, this necessitates critical thought and reflection on how we are structuring the spaces they're engaging. What is happening on the edges of active programming also informs youth education. Children absorb a vast amount of knowledge very quickly and from many different sources. This learning that takes place on the periphery is something we focus a lot of our attention to when we're considering the development of programming. Peripheral learning is constantly happening, whether that be in the context of how instructors are working together, how worldviews are being represented, what kinds of spaces we're gathering in, and how we can embody reciprocity and responsibility on a daily basis. To care for and learn with youth in the ways we wish, we must focus on the environments we are creating for them while they're with us. It is in this way that we co-create the context and conditions of optimal growth for the seeds of knowledge we intend to plant.

FOSTERING GROWTH AND RESILIENCY: OSHKWAZIN YOUTH LEADERSHIP PROGRAM

By Kristin Muskratt, Oshkwazin Youth Leadership Coordinator
TRACKS Oshkwazin is a program catered to high school-aged Indigenous students, focused on fostering leadership skills and building healthy relationships among youth. *Oshkwazin* is an Anishinaabemowin word, "invoking the idea of lighting one's fire, rising up and taking on the work we are meant to do" (Furgal, McTavish, & Smith, 2018). As the meaning suggests, we value youth's own agency in creating their path within the Oshkwazin Youth Leadership Program and beyond. We as educators can create spaces where youth can learn and grow

into leadership roles, but it is their individual decision to step up and choose to participate in these opportunities.

Oshkwazin is TRACKS' newest program and has been running active programming since 2018. The program aims to create a space where Indigenous youth can foster leadership skills, learn more about their own Indigenous cultural teachings and traditional ecological knowledge, while gaining practical and employable skills. All TRACKS programming focuses on braiding Indigenous and Western sciences. However, the Oshkwazin program particularly focuses on creating lasting and meaningful relationships between Indigenous peers, community, and mentors. We strive to create a safe and comfortable environment where youth can share any issues emerging from their own lives—from community or school life, including feelings of climate anxiety mentioned previously. By providing a space where Indigenous cultural teachings are shared and honoured, Oshkwazin offers the opportunity for youth to learn from one another and find solutions to the issues they are facing together.

I am proud to be the Coordinator of TRACKS Oshkwazin because I have lived experience that proves the importance of providing spaces for Indigenous youth to gain confidence and pride. Previously in my life, I have struggled with my confidence and Indigenous pride, and I wish I had been given more opportunity to bond with other Indigenous youth, learn traditional teachings, and meet Indigenous mentors.

Indigenous youth are resilient. However, it is unfortunate that we must speak to the resiliency of our youth as paramount. How do we provide spaces that allow Indigenous youth to thrive, instead of focusing on their resiliency? How do we foster the leadership skills that Indigenous youth will need to help future generations to combat environmental issues of today? TRACKS Oshkwazin seeks to do just that.

The three main components of TRACKS Oshkwazin are monthly lunch socials at local high schools, events such as youth summits and conferences, and our summer Youth Ambassador program. These Indigenous-only spaces are important as they create a safe, welcoming environment where youth can share life experiences and learn from one another through written and oral teachings. Through these experiences, we also connect youth with Indigenous mentorship opportunities that exemplify how Indigenous youth can thrive in a world that is stereotypically dominated by Western sciences and worldviews. We offer an opportunity for youth to realize that their ancestral and traditional ecological knowledge is scientific, and that there are spaces (i.e., workplaces, undergraduate programs, etc.) where Indigenous and Western sciences exist harmoniously.

Oshkwazin runs monthly lunch socials at a variety of schools throughout the Peterborough area and also connects with the Nogojiwanong Friendship Centre to run programming with its Wasa-Nabin Program, "a self-development program for urban, Indigenous, at-risk youth" (Nogojiwanong Friendship Centre, 2017). These socials and groups aim to build relationships between Indigenous youth living in Peterborough and in-community. Oshkwazin is filling a need in the community in this way. I have found through observation and personal experience that it is sometimes difficult for youth to connect because of their different life experiences. Urban and in-community Indigenous youth need a space to share and learn from each other as well. Through these groups, we can connect them with culture, knowledge sharing, and spirit.

We plan several events throughout the year that bring together Indigenous youth from across the Peterborough region. All Indigenous youth in the area are invited to attend these events, including those who have been previously involved with Oshkwazin lunch socials or youth groups. These events offer the opportunity for us to invite local Elders and knowledge holders to share teachings and facilitate workshops. These connections are integral to the program as they foster intergenerational learning, build relationships with potential mentors, and offer the opportunity to gain on-the-land teachings that can be shared with family and friends.

Indigenous youth also have the opportunity to apply for summer employment as an Oshkwazin Youth Ambassador. These positions have similar components to the lunch socials and larger group events but give the space for continuous learning and skills building throughout their summers. We connect to the land as much as possible during these ambassador experiences, and we encourage working and meeting outdoors. We also create a flexible schedule that can be changed depending on what teachings the land offers each day. This past summer, the youth ambassadors had a beautiful experience when they found serviceberries on one of our regular walking routes and received impromptu teachings that included stories and the harvesting of the berries. The youth offered semaa (tobacco) for the berries and spoke about the importance of reciprocity. There was a noticeable connection between the youth and the trees after this moment. For the rest of the summer, every time we walked by them, the youth would look or touch the trees, giving their silent hello and thank you. It is these lasting relationships with the land and with each other that we aim to foster through the Oshkwazin program.

SPREADING SEEDS

One of our past Oshkwazin Youth Ambassadors began his undergraduate degree at Trent University in September 2019. On his way to and from classes, he walks many of the same routes he did during the summer, and each time he walks past the serviceberry trees, he may remember his role in harvesting their berries and learning about the importance of reciprocity. By offering semaa before tasting the sweet fruit, the Oshkwazin Ambassadors engaged in a sacred cycle of reciprocity that we teach, and exemplify, at TRACKS.

Just as we learned with Beedahbin at Kinomaagewapkong, abundance is found in spaces where great care and attention has been put forward. As humans, it is our sacred responsibility to live in reciprocity with all other beings. When we honour this responsibility and learn from the land that supports us, we embody Anishinaabe aadiziwin. At TRACKS, we aim to embody these responsibilities and to teach others about the interconnected relationships between all species, natural and spiritual elements, and past and future generations. Through both the Outreach & Education program and the Oshkwazin Youth Leadership program, we put great care and attention into planting seeds and cultivating spaces that foster the growth of young leaders. As we enter our second decade of programming, TRACKS will continue to nurture the foundations upon which the program was built, honour our responsibilities to the land, and learn from all of our relations.

For more information on TRACKS Youth Program, visit www.tracksprog ram.ca

REFERENCES

Donald, D. (2012). Forts, colonial frontier logics, and Aboriginal-Canadian Relations. In A. A. Abdi (Ed.), *Decolonizing philosophies of education* (pp. 91–92). Boston, MA, USA: Sense Publishers.

Furgal, C., McTavish, K., & Smith, R. (2018). TRACKS youth program. *Pathways: The Ontario Journal of Outdoor Education, 30*(3), 17–21. Retrieved from https://www.coeo.org/wp-content/uploads/2019/05/Pathways_30_3.pdf

Hatcher, A., Bartlett, C., Marshall, A., & Marshall, M. (2009). Two-Eyed seeing in the classroom environments: Concepts, approaches, and challenges. *Canadian Journal of Science, Mathematics and Technology Education, 9*(3), 141–153.

Indigenous and Northern Affairs Canada. (2011). *Indigenous youth – Post-secondary education and the labour market*, based on the 2011 National Household Survey and the 2006 Census of Population. Ottawa, ON. Retrieved from: https://www.aadnc-aandc.gc.ca/eng/1451931236 633/1451932655379. Accessed October 23, 2020.

Nogojiwanong Friendship Centre. (2017). Retrieved from https://www.nogofc.ca/services/child ren-youth/wasa-nabin-program/

Nxumalo, F. (2019). *Decolonizing place in early childhood education*. New York, NY: Routledge.

Peltier, B. (2020, April 14). Wiikwemkoong first nation. *Personal Email Communication.*

Statistics Canada. (2016). *Aboriginal identity, STEM and BHASE (non-STEM) groupings*. Major Field of Study based on the 2016 Census. Ottawa, ON. Retrieved from: https://www12.stat can.gc.ca/census-recensement/2016/dp-pd/dt-td/Rp-eng.cfm? Accessed October 23, 2020.

Williams, D. (2018). *Michi Saagiig Nishnaabeg: This is our territory.* Winnipeg, Manitoba: ARP Books.

(Re)membering & Relationality

Family
active members of the universe

Elder
His words will forever stay

Creator
A single point of origin
-*nakt*- Time/space
Expanding,
Unfolding,

Mother Earth
She is a survivor

Past and present colonial practices
To think towards the future
Ethical relationality, A transactional form of imagination

Space and time relative to each other,
as opposed to being independent

Humans and non-humans
Co-exist

Ecological imaginations
A new avenue of possibilities
Multiple ways of coming to know celebrates balance and harmony

"Faces the wind"
A way for Elders to pass on knowledge to young people
Embodied Head (mind), Heart (emotion), and Hands (action)
Sharing stories connected to place
We move closer together

Kaa kishkaytaynaan taanishi lii Michif aen pimatishichik (We'll Learn About Métis Culture)

JOEL GRANT

My name is Joel Grant. I am a member of the Métis Nation of Alberta (Region 3). My immediate family consists of my father, my mother, and myself. I was raised in Cochrane, Alberta which is located in Treaty 7 territory. My outlet as a youngster happened to be sports; more specifically, I played competitive ice hockey. Playing hockey was a lot of fun. Upon entering junior high, I was accepted into a sports school on scholarship. It was here that I honed my skills because hockey development was incorporated into the curriculum. At age 14, I was selected as a member of the South Senators hockey team representing top bantam players from Southern Alberta. In subsequent years at the ages of 17 and 18 respectively, I played Junior "A" hockey (see Figure 8.1)_.

Figure 8.1. Grant playing for the Okotoks Oilers (Alberta Junior "A" Hockey League- AJHL) (2010).

Source: Author

Participating in team sports, including hockey, lends itself to experiences in leadership, collaboration to achieve goals, and learning how to accept challenges. I attribute these life lessons into my daily life.

As a young person, it was important for me to work hard in school. My parents were supportive and encouraged me to do well. My mother is Métis, and along with her brother they were the first ones to achieve university degrees in their family. It was always my mother's hope that I too, would pursue higher education. With much support from my parents, teachers, and coaches, I have been fortunate to have a good life and be able to avoid following a misguided path.

Figure 8.2. Grant's Mom & Dad on their honeymoon (in their 20's).
Source: Author

Our family name is Bonneau, and my mother and I are the only two Métis members remaining on my mother's side of the family (see Figure 8.2). We have suffered much loss to alcoholism, mental health problems, and poverty. While not forgetting any of these struggles regarding our history, it is important for me to have pride in being Métis, maybe because my mother always felt a sense of shame and guilt. When I was growing up, I always wondered why we never interacted with other Métis people. It wasn't until I was 13 years old that I met my mother's twin brother, Tim. Unfortunately, he passed away the following year. I found myself beginning to wonder where the rest of my Métis family was. Did I have other aunts and uncles? Where were my Métis grandparents? Do I have cousins somewhere? Although I am Métis, I felt very disconnected from our culture. Every day, I work hard to ease that sentiment and I now have a sense of pride and closeness to my Métis heritage that my mother never had.

There are reasons that are deeply rooted for the loss of culture in the Bonneau family. Their limited known history wasn't passed to her generation, and the rest of our family has died. My great grandparents from the Bonneau family passed away a long time ago—even my mother didn't meet them. My mother's Aunt, my Great Auntie Cecille, died in a psychiatric hospital in Ponoka, Alberta. My mother's Aunt, my Great Auntie Beatrice, passed away at age of 58 from alcoholism. My mother's only Uncle, my Great Uncle Frank, died from injuries suffered from a car accident. Also, my mother's favourite Aunt, my Great Auntie Dora, passed away in a Calgary nursing home from complications of diabetes. She was blind and very sickly when she passed. She was a great artist as she studied at the Banff School of Fine Arts. My mother had a close relationship with her and saw her on a regular basis. Lastly, my mom's mother, my grandmother Natalia (known as Tillie) raised twins. Unfortunately, she suffered from paranoid schizophrenia, making raising children more difficult. When she passed away at the age of 73, she was the last member of our extended family. I do not remember her because I was very young. From this extended family, there are some unknown cousins who I've never met. My mother has looked for them but has been unsuccessful.

My mother is an immense role model to me because she faced inconceivable adversity, growing up in and out of foster care, and yet, she has prevailed. She is a survivor. As a proud mother, she is incredibly passionate about building a strong sense of family. She has made a better life for herself and in turn, she has made a better life for me. I am extremely grateful for having her in my life. As I grew up focused on hockey and performing well in school, I never really asked my mom about her past; but she never offered to share her story. To this day, I don't feel comfortable questioning my mother about her hardships. It is my hope that she shares only what she feels comfortable with. Her story is her own and I am happy to be a part of it while we continue to navigate the intersecting truths of the past, the present, and the future. I am acutely aware that I grew up with love and support. Having two parents who spent a lot of time with me, has provided me with a sense of well-being.

Ghost River Rediscovery Society is an organization that assisted me in reconnecting with my culture. And in 2012 at the age of 19, it was an honour to be selected by the *Ghost River Rediscovery Society* of Alberta to be one of four Métis Youth Ambassadors to represent Canadian Métis youth. It was here where I gained a better understanding of what it means to be Métis. An opportunity to go to Tallinn, Estonia for six-months came about as an initiative coordinated through the *Métis Nation of Alberta* and *Rupertsland Institute*. In preparation for this meaningful excursion, I was invited to the *Ghost River Rediscovery Society* cabin just outside Calgary. There, we were greeted by a Métis Elder who shared her personal stories and experiences (see Figure 8.3). I was gifted a Métis sash and learned about the meaning behind the colours of the sash. This experience was dear to me because I felt that it was an awakening to my own identity, the beginning of my reconnection with my Métis roots.

While in Estonia, I was placed within the Non-Governmental Organization (NGO) *Continuous Action*, where I was a paid intern and attended Youth-in-Action training (aimed at intercultural exchange) in Turkey, Serbia, and Italy. But while working in the Estonian community with youths aged 13 to 17-years-old, I witnessed much drug and alcohol abuse. At the time, this was a global problem, and the issue was addressed at the many conferences I attended. I also worked at the

Tallinn food bank delivering hampers to those in need and helped children with disabilities at a local kindergarten. At the school, we assisted with mobility, lunch time assistance, and gym time. I learned the importance of helping others, which was significant to grasp at my young age. It is gratifying to help those in need or model behaviour for others, in hopes of instilling inspiration. It was at this time I applied to the Engineering Program at McGill University.

Figure 8.3. Grant sharing the Métis flag with Estonian youth. The Métis infinity symbol has a powerful meaning, representing the unity of French and First Nations Peoples, and the strength of Métis culture.

Source: Author

There was a time when I never dreamed I would be someone who would attend university, never mind having the opportunity to pursue a Master's Degree in engineering. I decided that engineering was an excellent way for me to develop my strengths in science and math while exploring opportunities offered by this exciting profession. Like many other young boys, I watched several television programs after school and hockey practice. One of my favourite shows was "Dexter's Laboratory." This show was about a mad scientist, boy genius named Dexter, conducting his experiments. I was riveted by the level of *coolness* with each science experiment. Dexter was building rocket ships, cloning machines, using complex algorithms to fulfill his mad scientist inclinations. As a high school student,

I remember a vivid discussion with my chemistry teacher about the discovery of the atom. Being that the father of nuclear physics, Ernest Rutherford was a professor at McGill, I think this inspired me to attend this Montreal institution. I often wonder what inspired me to pursue a degree in engineering. Not only did I have a curiosity about how things work, but I also found it fulfilling to solve scientific problems. I remember one of my homework assignments in high school where I had to calculate the physics of ice skating. I found this fascinating because I couldn't believe that my favourite sport could be explained with a few equations and technical drawings.

I choose to pursue a degree in Materials Engineering (with a minor in Biomedical Engineering) because it is a robust program with a wide range of applications. Materials science covers topics including steelmaking, nanomaterials, fluid mechanics, material microstructure, 3D-printing, additive manufacturing, material characterization, and biomaterials. This diverse program is an excellent starting point for building technical skillsets for problem solving. Young students can explore materials science applications in many fields of science. Learning about hip–joint and knee–joint replacements was the intersection between materials science and biomedical devices and showed me the complexity of material biocompatibility with the human body. Within a lab setting, I was able to contribute to a project of 3D-printing biocompatible hydrogels which has subsequently, been published in an academic journal.

Navigating the university environment was challenging. Early on, I struggled with achieving high grades. I remember the first day in my calculus class, we had a lecture that was basically high school review. It seemed like I had forgotten everything I had ever known about calculus, and I thought maybe this was going to be an uphill battle. But then I met my fellow classmates, everyone was talking about the credits they had received before entering McGill for "AP" and "IB" studies. I had never heard of "AP" and "IB." Apparently, there were specialized programs for *gifted* high schoolers that allowed them to get university credit for certain courses. I really felt like I was in over my head. I came from a public high school and played ice hockey. It seemed like I was lacking a sense of community where there was no support network. Feeling somewhat isolated, I had to dig in and self-reflect. How would I overcome these challenges? What needed to be done?

In 2015, in a winter semester co-op work term, I completed a four-month internship in Dhaka, Bangladesh. Dhaka is a city of nine million people. It was an incredible experience because I was stationed at a local university. There, I was tasked with writing a report on appropriate engineering technologies used in Bangladesh. While conducting a literature review on water filtration pumps and solar energy panels, I was impressed to see the work being done by an organization called Grameen Shakti. This organization creates renewable energy applications and sustainable health projects for the Bangladeshi people. Fortunately, I was able to visit their offices and operational engineering facilities to learn about the implementation of affordable technological solutions. The work they were doing changed my view of engineering applications. I saw a new avenue of possibilities I could pursue. Upon my return to Canada, I developed a deeper motivation to achieve higher academic success, but first, I would need a strong support network.

Upon the return from my internship in Bangladesh, I was looking for a place to live. As luck would have it, the coordinator at *First Peoples' House* (FPH) emailed me an offer to enter a new program whereby, Indigenous students at McGill could live at FPH. I only visited FPH a few times since arriving at McGill, such as welcome orientations and workshops. Little did I know that accepting the offer to move into FPH would turn out to be the best decision I would make during my undergraduate studies.

Before being at FPH, I focused everything I had into improving my learning and development in my engineering courses. But, FPH, was a place to connect with Indigenous students, mentors, and Elders, was exactly what I needed at that time. Living in Quebec, I was nervous about mentioning to anyone that I am Métis. In my experience, there was not a clear understanding of what it meant to be Métis from Western Canada to those who identify as being metis from Eastern Canada. There seemed to be deeply rooted misconceptions when I said I am Métis and from Alberta and I often found myself having to do a lot of explaining. I tried to make a clear distinction that Métis people are unique in that we have our own language(s) and culture. Many Métis people speak different dialects of Michif and engage in activities such as the Métis jig and fiddle. I do not (yet) know my language or how to jig, but it is my hope to reconnect and continue to learn as I navigate my lost culture.

Dwayne Donald's (2009) colonial frontier logics has allowed me to reflect on all my educational opportunities and experiences. While I am grateful for all the internships, hockey, and other learning opportunities, I wonder why learning about my language and cultures were largely missing throughout my schooling years. As I went through the learning process in the field of Engineering or playing hockey, I felt comfortable to acknowledge that I was building my expertise in these fields. However, I do not feel comfortable calling myself an expert in my own culture. I feel that my culture was deeply rooted within colonial frontier logics in both society and the education system, of which led to the sense of loss of my cultural values. When I was in school, the Alberta education system certainly skipped over Indigenous history and teachings. In recent years, there has been a lot of controversy because the Alberta Government has proposed removing the history of perpetrated violence towards Indigenous Peoples in Canada. Within this curriculum re-write, Alberta youth will also learn the bare minimum about

Indian Residential Schools, with a focus on European history and civilizations[1]. These actions are another representation of colonial frontier logics.

While at FPH, I was welcomed into the Medicine Bear Singers (see Figure 8.4). This is a group that drums at local ceremonies, Indigenous graduations, guest speaker events, and campus engagements. Every Wednesday, we would meet in the basement of FPH to practice songs shared by First Nations students from McGill. Even though I had no songs to share, I was eager to learn from our drum keeper who was from a Mi'kmaq community. He was open to sharing songs (not ceremonial) with a few Métis drummers, for us to learn and perform at events. We kept the songs to ourselves and practiced within our own community network at McGill. I would never share these songs because they are considered sacred to their own nations. The stories behind the songs have been passed down for generations and it would be disrespectful for me to share with others, especially since I am not Mi'kmaq. Singing has inspired me to learn about to perform the Métis jig, and drumming became an outlet for me to reduce my stress while attending classes.

Figure 8.4. Medicine Bear Singers getting ready to perform at an event at McGill.
Source: Author

Similar to the community support I received from my hockey development in my youth, I found myself receiving strong community support while at FPH. The close-knit community of Indigenous students, academics, and professionals created an incredible place for inspiration. It was a place to study, eat, and socialize.

We would host dinners and gatherings, and the FPH coordinator would facilitate monthly workshops such as beading, moccasin making, and job searching. It was a combination of cultural activities and professional development for Indigenous students. My grades were rising, and I was looking forward to the future.

The diverse backgrounds of Indigenous students' I met within the FPH community allowed me to realize that it doesn't matter where anyone is from or how deeply they are connected to their own culture. Everyone is accepting and supportive, and together we developed deep connections and build a strong community of support. Perhaps this is because we are a minority within the McGill community and many of us have experienced what it's like to negotiate our cultural values within an education system because of Colonial Frontier Logics (Donald, 2009).

There are many reasons why someone may not be connected to their culture. In the case of my family, there was a traumatic past. Unfortunately, this is too common among many Indigenous peoples. Instead of measuring and comparing one's connection to his or her own culture, I think that we as Indigenous peoples across Canada, need to unify. However, it is important to recognize the breadth of one's knowledge and connection to community. I am also very aware of the severity of people taking advantage of status and faking their Indigeneity. This is a common problem especially prevalent wIth people posing and claiming to be Métis. As I continue to navigate and reconnect with my lost culture, it is my hope that I can give back the same opportunity to other young Indigenous youth and children; much like how FPH has been a community for collective growth and support for myself and many others.

One of the most amazing experiences that I was introduced to while at FPH was the opportunity to be a Senior Camp Counselor for the *Eagle Spirit Science Futures Camp* (ESSFC) at McGill. The ESSFC camp promotes science and physical activity to Indigenous youth from across Canada. There is a major shortage of Indigenous doctors, engineers, and scientists (both at the urban and community levels). It was our hope that by offering spaces to appreciate both traditional knowledge and Western science, Indigenous youth will be inspired to pursue a profession in the field of Science, Technology, Engineering, Math, or Medicine (STEMM). It is inspiring to see how traditional knowledge from diverse Indigenous communities across Canada and Western science intersect. The ESSFC is held every year for one week in July for youth ages 13 to 17-years-old. The activities include morning science lab work, Elders storytelling, physical activity and sports, and visiting different faculties for tours and workshops. One of the facilities we toured was the Faculty of Medicine simulation center, where the youth had first-hand experiences of medical practice, such as stitching on mannequins and making molded denture samples. The youth then got to take these samples back home and share their experiences with others in their home communities. At another facility visit to the Faculty of Human Nutrition, we tried traditional cooking like chili and cornbread. These collaborative efforts between Indigenous and Western scientists, students, and Elders are essential for garnering Indigenous youth interest in STEMM. One example of passing on traditional knowledge was when an Elder shared an Anishinaabe Pike head teaching. Later in the day, we dissected Pike fish during a laboratory session. The kids become familiar with the parts of the fish with an understanding of both biology and traditional teachings. It was a great way to get the kids engaged while encouraging them to ask questions about creation stories in relation to science. It was in these moments, we hoped the importance of embracing both traditional culture and Western science was memorable for all. These two areas can, and should, co-exist and interact with each other, and provides meaningful reasons why young Indigenous youth can pursue higher learning by appreciating their own culture and knowledge beyond colonial frontier logics

I remember when we visited a Kanien'kehá:ka (Mohawk) Elder in Kahnawake. His words will forever stay with me. He mentioned that when he was young, he and his wife were excited to have children. However, they did not know their language or culture; it had been lost. They would not have anything to pass down from their parents. The Elder did not want his children to be like him and have no recollection of what it meant to be Mohawk. So, both him and his wife began to learn their language by reconnecting with the few Elders in their community. Throughout their house, they taped words on objects to identify their meaning in Mohawk. The whole kitchen and living room became covered with sticky notes containing Mohawk words for the objects they were placed on. This is how they practiced learning their language. Then he became an Elder himself and is now fluent in Mohawk. It took time and guidance but made them very happy. The Elder ended his message by portraying his belief that striving to reconnect and practice any one's lost language is the key to not forgetting those who have been silenced. He added, "Do not let it become a language of the past!" Through the story from this Elder, I have come to discover that I am learning and growing my own sense of why reconnecting with my culture is so important. It is not only for myself, but for my people and for future generations.

From September 2016 (until graduating), I have been the Vice President and President of the *American Indian Science and Engineering Society* (AISES) McGill Chapter (see Figure 8.5). Serving these positions included extensive volunteering commitments ranging from being an active member on the student Indigenous Perspectives Panel at the McGill First Peoples' House to planning community outreach events for promoting STEMM careers to Indigenous youth. In March of 2019, our chapter hosted the *National Canadian Indigenous Science and Engineering Society* (.caISES) conference at McGill.

Through my connection with .caISES, I am also a student member on the *Canadian Indigenous Advisory Council* (CIAC). CIAC is on the board of AISES and serves as the primary advisory group for Canadian Indigenous–STEMM promotion. I plan on being a member of .caISES for the rest of my life. For me, it is a community for collective growth and a great way for me to give back to other Indigenous youth through meaningful initiatives creating a platform to celebrate their own traditional knowledge in STEMM. I remember attending my first national conference in 2017 and being completely impressed by the size of the room hosting the welcoming event and the number of attendees. It was impressive to see over 2000 Indigenous Peoples involved with STEMM and an unforgettable experience. The network of AISES spans across North America. This is an

Figure 8.5. Grant attending various AISES National Conferences.

Top Left: Grant meeting commander John Harington (first Native American to fly into space).**Top Right:** Grant presenting a research poster and winning 3rd place at the 2017 AISES National Conference in Denver Colorado.

Bottom Left: Grant winning first place in the poster competition at the 2019 AISES National Conference in Milwaukee, Wisconsin.

Bottom Right: AISES McGill Chapter being awarded the Professional and Chapter Development Award at the 2019 AISES National Conference in Milwaukee, Wisconsin.

Source: Author

exciting time because it is a relatively new initiative in Canada, and I have been fortunate to be a part of CIAC during its novel development. The initiatives we focus on include sending our students to regional and national conferences, hosting speaker series events, and collaborating with the *Kahnawake Survival School* (KSS), one of the closest communities near McGill. We will continue to expand and build upon these projects. Being a member of .caISES gives me a greater sense of reconnection by being able to give back to youth and immerse myself in continued cultural learning.

As I conduct my research as a graduate student, I draw on many parallels with the teachings of the Métis sash. The Métis sash is full of many colours such as red, blue, green, white, yellow, and black. For example, green represents the fertility of the Nation is important and means the environment must be optimal for healthy growth. Red is the blood of Métis people. Blue represents the Métis spirit. White illustrates the connection to the Earth and our Creator. Yellow is hope for prosperity. Lastly, black is the suppression of peoples and dispossession of Métis land. The connection to the Earth and our Creator means that we must have a positive impact by making sure we are not polluting the Earth. By my youth experiences in sports, my traveling co-op work terms, and now attending university I am able to regain some of my lost culture. First Peoples' House (FPH) has been a beacon to lead me to other activities and social groups in my quest to reconnect. Also, writing this chapter was possible with the stories and reflection shared from my mother. I have come to realize the positive impact of having a strong support system that works. It is my hope that I have had a positive impact on others. Although I have not regained my culture fully, I am on my way.

NOTE

1 The Alberta Curriculum Re-write is an ongoing controversial issue, and at the time of this publication, has not been finalized.

REFERENCE

Donald, D. (2009). Forts, curriculum, and Indigenous Métissage: Imagining decolonization of Aboriginal-Canadian relations in educational contexts. *First Nations Perspectives*, 2(1), 1–24.

CHAPTER NINE

A Place-Conscious Approach to Teaching Mathematics for Spatial Justice: An Inquiry with/in Urban Parks

AMANDA FRITZLAN

INTRODUCTION

This is a study of teaching mathematics for spatial justice with urban middle school students in North Vancouver, British Columbia. It takes the form of a proposed series of lessons that inquire into the right for all people to have access to natural outdoor places such as parks. This educational endeavour is attentive to the diversity of students' cultural and political experiences, histories, and engages in a conversation with Donald's (2012) concept of ethical relationality, "a transactional form of imagination that asks us to see ourselves implicated in the lives of others not normally considered to be our relatives" (p. 93).

Spatial justice bridges issues of social justice and critical thinking about human relationships with places. Theories of spatial justice recognize that "social justice cannot abstractly be reached, since social relations take place in a particular space" (Watson, 2020, p. 1). The spatial justice issue that this study engages with is access to urban designated park land. It is inspired by my experiences of working for over a decade with middle school students living in a dense urban area. A part of my teaching practice has been to facilitate students' access to "natural" outdoor places. Often, this has meant walking long distances with students to find parks and forests. Access to parks or "natural" outdoor places as an issue of international human rights[1] is also considered in this study.

A Platonist notion of mathematics is characterized as "being value-free, decontextualized, a cultural, non-ideological, purely objective in its use, always consistent, generalizable, universalist in the sense of being universally true" (Aikenhead, 2017, p. 27). This abstract conception of mathematics traditionally privileged in Canadian public schools offers quantifiable descriptors for a place and for relationships between places including: length, height, width, area, volume, angle, and circumference. Abstract spatial reasoning may be criticized as having "no connections to social dynamics between people or power relations that structure lived-in spaces" (Rubel, Hall-Wieckert, & Lim, 2016, p. 556). In response, this study adopts a notion of mathematical thinking and practices as a form of social relations (Rubel & Nicol, 2020, p. 173). This study works with a broader sense of mathematics that takes into consideration Bishop's (1988) six aspects of culturally linked mathematical thinking: counting, measuring, locating, designing, playing, and explaining. Particular attention is given to how the ways of expressing mathematical thinking mediate students' relationships to a particular place.

For the purpose of this study, a "particular space" will be referred to as a "place". Acknowledging social, cultural, and historical meanings connected with particular places is central to this mathematics for spatial justice inquiry. Examples of this are considering the names of places, stories connected to places, and students' experiential and cultural histories in relation to places. Paying attention to power relationships or questions of spatial justice in mathematics education goes beyond simply linking the social to place; it links mathematics education to place-conscious pedagogies that consider what has happened, what is happening, and what is appropriate to have happen in a particular place (Greenwood, 2013).

This inquiry focuses on middle school education in North Vancouver, British Columbia, the traditional, ancestral, and unceded territories of the Squamish and Tsleil-Waututh Nations. The historical and present-day ways that these Indigenous communities relate with place are necessarily a part of a spatial justice inquiry in this place. Donald's notion of ethical relationality (2016) addresses the complex interface between worldviews and experiences of local Indigenous peoples, municipal governance of the parks, and the students who have access to the parks.

This inquiry asks: In what ways can mathematics education engage with complex aspects of spatial justice in relation to urban parks? It develops a series of lessons for raising consciousness and empowering students. The methodology for this study is inspired by the work of Rubel, Hall-Wieckert, and Lim (2016) to present a "design heuristic" to teach mathematics for spatial justice that expands on previous experiences of educators, students, and researchers in particular places. The series of lessons that I propose are arranged within a framework of three pedagogical approaches: a numerical data approach; a place-conscious approach; and a class discussion approach. For the design of these lessons, I draw on my experiences as a middle school teacher in North Vancouver, BC with a culturally diverse community of students. While this inquiry is specific to a particular place and community it contributes to larger conversations of mathematics education for social justice.

MATHEMATICS EDUCATION FOR SPATIAL JUSTICE

Theories of critical mathematics are key to understanding mathematics education for spatial justice. Gutstein (2003) explains that critical mathematics involves both "reading the world using mathematics" (p. 48) and also "writing the world" with mathematics (p. 40). Injustices and inequities may be understood and articulated using mathematics. As well, mathematics may be used as a tool to challenge those same issues. Similarly, a spatial justice approach to mathematics education considers "the importance of providing a balance between, on the one hand, opportunities to recognize oppression and, on the other hand, opportunities to exercise reimagination or action toward transformation" (Rubel et al., 2016, p. 557).

Theories of critical mathematics and spatial justice are both concerned with issues of social justice. Integrating these theories brings together critical mathematical thinking and practices with an acknowledgment of social relations and issues of injustice and inequities emerging and manifesting in relationship to particular places. Soja (2010) theorizes spatial justice as: "everything that is social (justice included) is simultaneously and inherently spatial, just as everything spatial, at least with regard to the human world, is simultaneously and inherently socialized" (p. 24).

As is illustrated in the middle school lessons described below, abstract measurements of space may be used as a tool for comparing equitable access to places that are assumed to be similar, such as municipal parks. Decoding numerical data of spatial human relationships gives students the tools to engage in critical conversation of place and power relations. Building on mathematical understanding of dominant representations of place empowers students to challenge and re-math their worlds.

Working with maps and mapping can be a significant part of mathematics education for spatial justice. By understanding maps as non-neutral, arguing for a specific way of imagining space, representing specific points of view, and making visible certain entities or patterns, teachers and students can engage simultaneously with the spatial and the social justice aspects of cartography. Introducing students to alternate ways of conceiving of place can uncover status quo rhetorical and technological aspects of representations with maps. Rubel, Hall-Wieckert, & Lim (2016) describe "participatory mapping" exercises in which students create maps based on their own interactions and interpretations of a place as a way of "challenging existing hegemonies" (p. 560).

PLACE-CONSCIOUS EDUCATION

Along with theories of critical mathematics and spatial justice, the framework for this study includes a theory of place-conscious education (Greenwood, 2013). In public school settings, place-based education has historically endeavoured to make connections between students and local natural environments through physical, hands-on activities, and knowledge of local plants and animals (Orr, 1992; Sobel, 2004). A fundamental difference between place-based education and a critical pedagogy of place such as Greenwood's (2013) place-conscious theory is the discussion of social, historical, epistemological, and ontological connections with place.

A "place-conscious" approach to education guides the design of lessons for this study in the context of urban parks. Students' are encouraged to think beyond the immediate municipal public presentation of parks, for example in the form of signage, names, plants, wildlife, and built structures. Considering the history of the places that we visit is significant and affects interpretation of what is happening in the parks as we spend time there. Students' work to reimagine and remake parks and the surrounding places through activities such as making their own maps aligns with Greenwood's "place-conscious" imperative to ask what is appropriate to have happen next in place. Writing the world with mathematics for social justice can take many forms. Countermapping, participant mapping, and critical examination of municipal statistics for citizen access to parks are a few.

Humans and non-humans shape place through physical interactions: birds choose flight paths, insects' nest, rain pools, and a person plants a tree. As well, communities of humans shape place through creating narratives and upholding of those narratives, such as when a city council votes on the designation of a particular place for use as a park and the name that will go on the signage and municipal maps. Thinking critically about social narratives in place, opens discussions about power relationships and agency.

"Place acts as a shaping force in social relations, whether among humans or non-humans" (Rubel & Nicol, 2020, p. 175). Considering mathematics as a practice of social relations means that place shapes mathematics. Mathematical thinking and practice, conceived as a social relation, takes the form of a diversity of cultural practices that relate with place. Culturally based ways of relating with particular places make up the practices that Bishop (1988) identifies as mathematical through his six universals for culturally linked mathematical practices (counting, measuring, locating, designing, playing, explaining). Working with students to record qualitative data for their own cultural experiences and histories in relation to a place during lessons that facilitate personal writing, storytelling, or artwork engages with issues of diversity of social values and meanings of place. Teaching in/with place creates opportunities to shift our social relations as teachers and learners.

COLONIAL FRONTIER LOGICS AND ETHICAL RELATIONALITY IN EDUCATION

Donald (2012) articulates a "colonial frontier logics" of Aboriginal-Canadian relations that applies to education. He writes, "The overriding assumption at work in this logic is that Aboriginal peoples and Canadians inhabit separate realities" and he points to an embedded agenda to "deny relationality" (p. 91). Within this system, the fort is seen as a symbol of Canadian nationhood, and at the same time, of protection, as a step in civilizing Canadian wilderness.

Donald (2012) argues that the mythology of the Canadian fort as bringing civilization to the wilderness followed by progressive development can be extended to other Canadian institutions including schools (p. 100). Recognizing the colonial frontier logics operating in educational institutions exposes existing power dynamics. Schools as Canadian institutions in this system of colonial frontier logics (including classrooms, policies, curricular documents, and governing structures) create a separation between Indigenous and settler peoples. As such, a "school, classroom, or curriculum document, is received as a fait accompli wherein significance and relevance is already decided". In this way, "aboriginal perspectives and knowledge systems in educational settings are discounted from the outset" (Donald, 2012, p. 101).

The decision within this study to take some of the activities outside, into a park, is an action of stepping away from the physical school building symbol of colonial frontier logic and fort mythology in education. However, curricular direction, and other mandated or assumed teaching and learning structures are still present. Additionally, designation of land for controlled park use that I move away from the school building to visit with students may also be considered a manifestation colonial fort mentality.

Inviting and acknowledging the multiplicity of students' perspectives of being in parks within the proposed lessons of this inquiry has the potential to counter colonial frontier logics through a practice of ethical relationality. Donald (2012) explains that ethical relationality "requires ecological imagination to be enacted and repeatedly renewed. This is a transactional form of imagination that asks us to see ourselves implicated in the lives of others not normally considered to be our relatives" (p. 93).

Donald's idea of ethical relationality aligns with work by Dion (2008) to challenge what she identifies as a "perfect stranger" position adopted by individuals that claim to have no relationship with Indigenous peoples in Canada. Dion works with individuals and groups of educators to understand their own relationships and responsibilities to Indigenous cultures, identities, and histories. A study of place that includes Indigenous stories and philosophies as well as past and present colonial practices follows a path of awakening students to a self-knowledge of their own positions or roles in that place.

LESSONS DESIGNED FOR TEACHING MATHEMATICS FOR SPATIAL JUSTICE WITH/IN URBAN PARKS

1. A Numerical Data Approach

This set of lessons for teaching mathematics for spatial justice with/in urban parks begins with gathering online government data for multiple cities around the world including: total area, total area of parks, populations; and area of park per individual. Presenting a table, as seen in Table 1 below, to the students in varying degrees of completion emphasizes particular mathematical skills and caters to levelled ability.

City	Area (km^2)	Population	Density (people/ha)	Park area (ha)	Park area %	Park area/person (m^2/person)
Vancouver[2]	114	631 486		1254	11	20
Toronto						
Tehran						
Nairobi						
Singapore						
?						

Table 9.1. Gathering and Calculating Statistics for Urban Populations and Park Area.

Source: Author

Data for this exercise is gathered by the students online individually or in groups and is partially given by the teacher. Students add different cities where they may have lived or traveled. Relationships between the different data columns provides for practice with several numeracy skills. For example, familiarity with and conversion between metric units for area are necessary for this mathematical exercise. There are three different units used on the chart: km^2, ha, and m^2. A discussion of why these units were selected and whether it would be better to have numerical values represented all in the same unit furthers students' comprehension of the units. A visual representation of these units on a local map is also informative. Operations with decimals and percentages are mathematical skills required to complete the Table 9.1 exercise. If different units are used for area, rounding off to a consistent number of decimal places needs to be instructed or negotiated. Understanding the fact that online sources may also round off numerical representations of land area or populations of people is pertinent to these exercises.

Comparison of percentage of total park area for different cities can be interpreted as indicating the importance of access to parks for people in particular cities. However, considering municipal data expressed as ratio of area of park per individual person in a particular city adds another dimension for considerations of park access. Once the students' tables have been completed, the following questions develop students' numerical literacy skills and also invite students to critically think about the relationships between abstract representation of parks and lived experiences:

1. Does a high percentage of park area correspond to a large park area/individual person?
2. Is a large park area/person always more advantageous than a small park area/person? Give an example from your own experience to support your answer
3. What doesn't this table tell us about people and park use in urban centres?

The next part in this numerical comparative approach to inquiry of parks in cities is to have students create a full scale outdoor interactive representation of park area/person for a city of their choosing. As I was designing these lessons, I decided to try this myself. I chose the city of Vancouver, where I live. The numerical data for area of park/individual person, I figured out to be 20 m². I measured this out in a park close to my home with some twine and sticks and then stood inside the 2 m by 10 m container that I had created. I invited a friend to stand inside the twine boundary with me. I found that standing there felt limiting to myself and towards others whom I could see from where I was standing. Would I have felt different if there had been a tree in this particular 20 m²? My friend commented that this delineated area was not useful except for maybe lying down or stretching. People walking by looked at us, confused at what we were doing, but not intrigued enough to stop and watch or ask. A dog ran under the twine.

This active component of the inquiry takes place outside. It is another way for students to critically engage with abstract mathematical concepts in relation to their own physical worlds. The information represented in Table 9.1 gives an indication of park land available to people in different cities, however, it gives no indication of the way in which people use or relate to particular parks. This embodied mathematical exercise has the potential to expose the limitations of mathematical language to articulate how we relate with places as it excludes movement, spatial sensory experiences, and social interaction.

2. A Place-Conscious Approach

The following lessons consider the notion of access to urban parks through a walking field trip to a park. Within the classroom, a city map may be used to plan the field trip. Employing cartographic tools of scale and a Cartesian grid, students can map and measure the route. Leaving the classroom, maps in hand, students will be prepared to follow the prescribed route. Noticing differences between the information that the map provides and actual experiences of walking to a particular park provides a critical look at the use of grids in location and navigation. Doolittle (2018) writes of the failure of Cartesian grids to create maps that match our physical world. He gives examples of the failure of a Cartesian grid in attempts to create straight roads in Saskatchewan, as the earth's spherical properties challenge the two-dimensional maps, and in the Niagara Escarpment's cliff-like edges being hidden by the grid road map for Hamilton, Ontario.

As students walk, they may notice concrete, gravel, and grass surfaces that they step on, as well as shade, wind, and changes in elevation that affect their experience. Students can also be asked to take note of the green places that are not designated as parks, the liminal in-between places (Gerofsky, 2018) inhabited by plants, wildlife, and waterways. Asking students to consider what some of the other pathways through this area are, for example, those of the birds, insects, and wind, further disrupts the authority of one kind of map.

Once the students have reached the park that the class set out to find, they may practice using alternate modes of "locating" using their senses to describe where they are, perhaps with eyes closed, or observing with five senses and writing it down. A map may be collectively created using senses, writing stories in place, and taking photographs. This process of participatory mapping (Rubel, Hall-Wieckert, & Lim, 2016, p. 560) shapes the meaning of this park to the students and is responsive to their individual and collective experiences and histories.

Connecting narratives with place is a way of understanding socio-historical meanings of place and of examining the ways in which the usage of urban park may or may not be socially appropriate. Attention to histories, oral narratives, as well as particular landmarks and how places are named, toponymy, are ways of acknowledging with students the interconnectedness of language, culture, and the physical world (Basso, 1996).

One of the parks that I have taught with/in is Whey-ah-Wichen. Whey-ah-Wichen is a Tsleil-Waututh name which means "faces the wind". It belongs to a rounded point of land that protrudes into or is the corner of a bend that I know as the Burrard Inlet. Wind coming from the Salish Sea, also called the Georgia

Straight, to the west, can be felt if you stand on the shore. Maureen Thomas, Tsleil-Waututh Elected Chief explains, "The naming of places is an important part of our cultural identity and signifies for Tsleil-Waututh the connection we have with our land." In 1950, the local municipal government gave the name Cate's Park to Whey-ah-Wichen in memory of Charles H. ("Cates Park, Deep Cove, BC", n.d.). Cates arrived in Vancouver in 1886 and set up a tugboat company, C. H. Cates and Sons. This business continued to grow and was amalgamated into Seaspan in 1999 (Seaspan, 2016). The traditional Tsleil-Waututh name, Whey-ah-Wichen, was restored in 2001("Cates Park, Deep Cove, BC", n.d.).

Another example of the connection between names and meanings of places is the Squamish name for twin mountain peaks visible from many parts of North Vancouver is Schi'ich'iyúy, which translates to "The Sisters" in English (Nelson-Moody, George, & Joseph, 2009, p. 81). European settlers renamed these peaks "The Lions" in reference to the "Landseer Lions in Trafalgar Square ... splendid sculptures ... of Old England" (Johnson, 1912, in Johnson & Butt, 2016). In an illustrated storybook (Johnson & Butt, 2016) version of The Two Sisters, a lesson of people learning to live peacefully together in place is linked to these side-by-side peaks:

> And on the mountain crest the Chief's daughters can be seen wrapped in the suns, the snows, the stars of all seasons, for they have stood in this high place for thousands of years, and will stand for thousands of years to come, guarding the peace of the Pacific Coast and the quiet of the Capilano Canyon. (Johnson & Butt, 2016)

Archibald (2016) writes of the way that Indigenous storywork has traditionally been a way for Elders to pass on knowledge to young people and that it has been brought into schools to follow the young people. Principles of Indigenous storywork include "respect, responsibility, reverence, reciprocity, holism, interrelatedness, and synergy" (p. xi). Protocols for storywork guide storytellers to respect who a story belongs to and who may tell a story, where and when is an important consideration of these lessons with middle school students.

Another narrative of the parks that students may visit in North Vancouver is that it occupies traditional ancestral unceded territories of the Squamish and Tsleil-Waututh Nations. Ontologies of land ownership and usage surface in a place-conscious inquiry of a specific place that considers Indigenous and colonial histories. Private land ownership was not a part of Tsleil-Waututh or Squamish culture prior to contact with Europeans. "First Nations did not have boundaries between them, but operated on a system of protocols and customs which were known and respected by everyone and allowed access to each other's resources" (Tsleil-Waututh Nation, 2016). This is a philosophical and political shift for most non-Indigenous students. Even though the homes of many students and members of the school communities that I work with have homes on local reserves, there may be many students who do not know that reserve systems were imposed on Indigenous people and that the Tsleil-Waututh and Squamish Nations have never ceded their territories. Including this conversation in a study of spatial justice that focuses on access to parks challenges not only the systems by which parks are created, but also the notion of access. Related lessons in mathematics for spatial justice could centre on Indigenous territories and treaties in Canada.

3. A Class Discussion Approach

The following questions are meant to provide starting points and structure for class discussions about the concepts of park and of place after having participated in the lessons of the previous two sections:

1. What is a park? Why are there parks in cities?
2. What is the history of the parks that we spend time in? What was this place before it was an official park?
3. Does everyone have a right to access parks? Are all park areas accessible to all people in a city? How can we find this out? How much park area do we each need?
4. What can the language of mathematics and numbers help us to understand about outdoor places? What can't it tell us?
5. Do all places have meaning? Do they mean the same thing for all people? Can the meaning of a place change?
6. Are there other natural places besides parks in a city?

Including mathematical language and concepts in these reflections allows for critical discussion of its use and meaning. Students may notice how their combined numerical and experiential approaches to inquiring about parks has shifted their perspectives and understanding. A place-conscious pedagogy applied to parks can extend the ability of students to think critically about other places in relation to their socio-cultural worlds.

Many of the middle school students that I have worked with had recently immigrated to Canada and I knew little of their personal and cultural histories of accessing parks. What were their experiences of urban parks? What did they expect to do and see in parks? Giving the students the opportunity to express their experiences and perspectives opened a conversation of how we relate to parks. In one such conversation, students shared their knowledge that: in Nairobi, Kenya, many parks serve as places for recreation, wedding photos, and to remember historical events; in Tehran, Iran the primary use of parks is for families and access to nature; and in Shanghai, China air pollution prohibits access to parks on some days and there is not enough access for individuals with disability. This qualitative information that compares urban parks in different parts of the world illustrates for students the difference in the defining features of urban parks in terms of access, use, civic identity, and meaning.

Listening to stories, histories, and multiple perspectives can trouble colonial frontier logics embedded in educational institutions in Canada that continue to divide people and ideas. Through exploring cultural differences together as a learning community and acknowledging being in a particular place together, students with diverse personal and cultural histories may realize that they are implicated in one another's worlds, in other words, enacting ethical relationality:

> This concept of relationality instantiates an ethical imperative to acknowledge and honour the significance of the relationships we have with others, how our histories and experiences position us in relation to each other, and how our futures as people in the world are tied together. (Donald, 2012, p. 104)

CONCLUSIONS

The use of numerical data to describe equitable park access serves large model descriptors and comparisons. The tensions between the particular and the universal in local versus global studies is a dimension of social and spatial justice that requires constant negotiation. An inquiry into urban parks relates to local histories and individual student experiences. Connecting this conversation to the language of mathematics empowers students to use the tools of mathematics in ethical relationality to place and to others in their communities. Teaching mathematics for spatial justice extends "the critical mathematics goals of reading and writing the world to include spatial perspectives. In other words, reading the world with mathematics suggests sociopolitical consciousness about social factors that impact the development of place and space" (Rubel, Hall-Wieckert, & Lim, 2016, p. 562).

In Canada, sharing stories connected to place, particularly focusing on the traditional stories of local Indigenous students and their communities, upsets colonializing assumptions that the land was empty, terra nullius, before European settlers arrived. Indigenous peoples' ways of perspectives and stories of place broaden all students' understanding of the meaning of places.

> Sustained attentiveness to Aboriginal-Canadian relations and willingness to hold differing philosophies and worldviews in tension creates the possibility for more meaningful talk on shared educational interests and initiatives. (Donald, 2012, p. 107)

When students experience connection to a particular place, their understanding takes on embodied phenomenological and historically conscious dimensions. The transformative possibilities of this study centre around how we as educators and as students think, feel, and talk about a particular place as well as how we envision spatial justice using mathematics.

My understanding of the use of numbers to represent human use, presence on, or connection to parks has shifted through this study. Numeric ratios applied to parks and people, for example 20 m² of park land/person in Vancouver, assumes that all parks and all people are the same. This approach does not consider the topographic, social, historical, and cultural significance of a particular place or the relationship between people and place-based on their own group and individual cultural experiences. Attention to relationship with place, cultural perspectives, and worldviews expands the meaning of parks as being accessible for humans. An extension of this inquiry considers other than humans who have globally varying legal rights (Newearth, 2016; Yale Centre for Environmental Rights and Policy, 2015).

How does this method of using multiple approaches to teaching about urban parks bring into question the definition of right to access parks? What if it was a right to culturally responsive land education? Or, the right to relationship with land in accordance with cultural worldviews? How might mathematics be used to promote a right to culturally appropriate relationship with land? The concept of spatial justice contributes to a conversation of finding balance between identifying oppression and reimagining or taking action for transformation in the context of mathematics education (Rubel, Hall-Wieckert, & Lim, 2016). Working with the tensions and connections that arise from telling and listening to all of our stories and histories in a shared place, such as an urban park, follows Donald's principle of ethical relationality in education: "Through the reciprocal process of teaching and learning, we move closer together" (Donald, 2012 p. 102).

NOTE

1   This paper works with the assumption that all human beings have the right to connect with the land that is in a reasonably natural state. However, this could be a topic of research and debate for a classroom as an extension exercise. There are online documents that support international rights to nature for the child: https://www.ohchr.org/_layouts/15/WopiFrame.aspx?source doc=/Documents/HRBodies/CRC/Discussions/2016/DGDoutcomereportMay2017.docx-&action=default&DefaultItemOpen=1

REFERENCES

Aikenhead, G. S. (2017). *School mathematics for reconciliation: From a 19th to a 21st century curriculum.* Retrieved from https://www.usask.ca/education/documents/profiles/aikenhead/index.htm

Basso, K. H. (1996). Wisdom sits in places: Landscape and language among the Western Apache. Albuquerque, NM: University of New Mexico Press.

Bishop, A. J. (1988). Mathematics education in its cultural context. *Educational Studies in Mathematics, 19*(2), 179–191.

Cates Park, Deep Cove, BC. (n.d.). Retrieved from http://www.deepcovebc.com/deepcovebcpark strails/cates-park/

Dion, S. (2008. Disrupting molded images: Identities, responsibilities and relationships – Teachers and indigenous subject material. *Teaching Education, 18*(4), 329–342.

Donald, D. (2012). Forts, colonial frontier logics, and Aboriginal-Canadian relations: imagining decolonizing educational philosophies in Canadian contexts. In A. A. Abdi (Ed.), *Decolonizing philosophies of education* (pp. 91–111). Rotterdam, ND: Sense Publishers.

Doolittle, E. (2018). Off the grid. In S. Gerofsky (Ed.), *Contemporary environmental and mathematical education modelling using new geometric approaches* (pp. 101–121). Cham, Switzerland: Palgrave Pivot.

Gerofsky, S. (2018). Always an abundance: Interstitial/liminal space, time, and resources that are invisible to the grid. In S. Gerofsky (Ed.), *Contemporary environmental and mathematical education modelling using new geometric approaches* (pp. 47–65). Cham, Switzerland: Palgrave Pivot.

Greenwood, D. (2013). A critical theory of place-conscious education. In R. Stevenson, M. Brody, J. Dillon, & A. Wals (Eds.), *International handbook of research on environmental education* (pp. 93–100). New York: Routledge. doi:10.4324/9780203813331.

Gutstein, E. (2003). Teaching and learning mathematics for social justice in an urban Latino school. *Journal for Research in Mathematics Education, 34*(1), 37–73.

Johnson, P. (1912). *Legends of Vancouver* (2nd ed.). Vancouver, BC: Geo S. Forsythe & Co.

Johnson, P. (Tekahionwake) (Author), & Butt, S. (Illustrator). (2016). *The two sisters.* Vancouver, BC: Waterlea Books.

Nelson-Moody, A., George, G., & Joseph, T. (2009). *People of the land: Legends of the four host nations.* Penticton, BC: Theytus Books.

Newearth. (February 16, 2016). *Bolivia passes "Law of Mother Earth".* Retrieved from https://newea rth.media/bolivia-passes-law-of-mother-earth/

Orr, D. (1992). *Ecological literacy: Education and transition to a postmodern world.* Albany, NY: State University of New York Press.

Rubel, L. H., Hall-Wieckert, M., & Lim, V. L. (2016). Teaching mathematics for spatial justice: Beyond a victory narrative. *Harvard Educational Review, 86*(4), 556–579.

Rubel, L. H., & Nicol, C. (2020). The power of place: Spatializing critical mathematics education. *Mathematical Thinking and Learning, 22*(3), 173–194. DOI: 10.1080/10986065.2020.1709938

Seaspan. (2016). C.H. *Cates and Sons.* Retrieved from http://www.seaspan.com/history

Sobel, D. (2004). *Place-based education: Connecting classrooms and communities.* Great Barrington, MA: The Orion Society.

Soja, E. (2010). *Seeking spatial justice.* Minneapolis-Saint Paul, MN: University of Minnesota Press.

Tsleil-Waututh Nation (2016). Tsleil-Waututh people of the Inlet. Retrieved fromhttp://www. twnation.ca/About%20TWN.aspx

Yale Centre for Environment Law & Policy. (2015). *The politics of rights of nature in Ecuador: Natalia Greene.* Retrieved from https://vimeo.com/119182093

Relationship-Based Science Education: Understanding the Mother Earth Through the Engagement of Head, Heart and Hands Through Artful-scientific Inquiry

EUN-JI AMY KIM
RO'NIKONHKÁTSTE NORTON
HANNAH KARAHKWENHAWE STACEY

Ms. Kim introduced local Indigenous peoples' practices in her grade 10 science classroom in the city of Montreal. One of her students shouted out: "Miss, why do we have to learn this in science? This is History!" The student's question raised questions central to this chapter: What made this student not be able to see the place of Indigenous knowledge in science classroom? Also, in what ways teachers can challenge students' biases and assumptions on what "science" should be?

The term "science" originally came from the Latin word, "*scientia*", which means "knowledge in the broadest possible sense" and science simply meant knowledge in archaic English (Snively & Corsiglia, 2001, pp. 8–9). Contrast to its original meaning, science, as an academic discipline now has a narrow focus on "Western Modern Science (WMS)" (i.e., Eurocentric Science) since British Association for the Advancement of Science (BAAS) deliberately used the term to denote their own knowledge and process as "science" in the 19th century. Since then, science education has become a place where WMS is considered the only form of valid knowledge while non-WMS content is treated as "cultural" or in the case of Ms. Kim's class, "historical" content, despite that the origin of WMS goes back to the ancient Greek, Egyptian and Mesopotamian philosophers. Such a hegemonic view of science education is driven by *colonial frontier logics* (Donald, 2009).

Dwayne Donald (2009), a Papaschase Cree scholar, describes colonial frontier logics as the "epistemological assumptions and presuppositions, derived from the colonial project of dividing the world according to racial and cultural categorizations" (p. 20). Colonial frontier logics separates people from the land, and people from the people, continuing the compartmentalization of knowledges and division of peoples. It makes us forget the relationship one has with another as well as with the Land, thus fails to acknowledge the "rationality and connectivity that comes from living together in a place for a long time" (Donald, 2009, p. 6). Sadly, colonial frontier logics are deeply embedded in the cultures of many educational institutions. In turn, the status quo between Indigenous knowledge and Western knowledge (i.e., Western Modern Science) perpetuates intended or not. Coupled with neoliberal globalization, colonial frontier logics work to hide the importance of connectivity and relationships in the field of STEM and science education.

Neoliberalism also places science as a discipline as an elevated status in society. Meritocracy is a key neoliberal ideology promoted in the education system, as can be seen in the current worldwide move to adopt standardized testing to promote individual educational and wealth attainment. As well, the content (i.e., knowledge and practice) studied to prepare for such standardized testing is introduced in science classrooms as objective knowledge, and as such is understood as universal and the truth. In this light, students must master the skills, knowledge, and ideologies promoted in their science classrooms in order to succeed in STEM driven societies. In such a setting, content that is "objective"—which can be measurable and comparable has a higher place than "subjective" content which promotes diverse ways of coming to know and multiplication of knowledges.

We view science as the inquiry of coming to know nature (Mother Earth) and our relationship with one another. In this context, teaching and learning science involves one's 3H: Head (mind), Heart (emotion), and Hands (action) (Kim, 2018). Tanaka (2016) suggested that "Indigenous teaching focuses as much as on 'learning with the heart' as it does on 'learning with the mind'" (p. xii). She also mentioned that "using 'good hands' by having a clear mind and healthy intent are deepened through a focus on physicality and doing" (Tanaka, 2016, pp. 22–23). As such, learning/teaching science involving 3H is a type of "embodied encounters" that includes "emotions, cultures, [and] physical sensations" (Derry, 2005, p. 35). In such embodied learning science involves practices of Art, not as a "hook" for engaging learners but as an inquiry process for a research: generating research/inquiry questions, collecting, processing, analyzing information, communicating ideas. Meanwhile, Shawn Wilson (2008), Cree scholar, mentioned that "research is ceremony ... [and] the purpose of ceremony is to build stronger relationships, or bridge the distance between aspects of our cosmos and ourselves" (p. 11). In this light, we use the term, artful-scientific research to describe our coming to know process that appreciates multiple ways of knowing—both Indigenous knowledge & Western Modern Science and/or Science & Arts as disciplines.

We are three educators and educators in training met in Kahnawake, Mohawk territory during the course of elementary teaching science methods offered by Kahnawake Education Center-McGill University. In the course, we had an opportunity to delve into artful-scientific research project that involves 3H. Our entry points and academic and cultural backgrounds may be different, but we all situate ourselves as learners and committed to engage in lifelong learning processes through multiple ways of coming to know.

Ro'nikonhkátste Norton is a Kanien'kéha language learner, instructor, and advocate. He currently teaches Kanien'kehá:ka at Kanien'kéha Ratiwennahní:rats Adult Language Immersion Program.

Hannah Karahkwenhawe Stacey is a Mohawk woman from Kahnawake. She is a pre-service teacher enrolled in the B.Ed program at Kahnawake Education Center-McGill University. In between classes, she tutors elementary and high school students in the community in Math, Science, English, and French. In her spare time, she is usually found reading a fantasy or mystery novel.

Eun-Ji Amy Kim is a first-generation settler from Korea. She considered herself to be in the process of "becoming an ally" to Indigenous peoples around the world, thus a lifelong learner (Bishop, 2015). A former high school teacher, she now works as a science curriculum consultant at Kahnawà:ke Education Center.

What follows is our individual stories of coming to know through a process of engaging with this assignment. Here, we would like to note that we are not sharing the tips of teaching/learning Indigenous (Onkwehón:we) knowledges in science. In coming to know Onkwehón:we ways of understanding Mother Earth is a lifelong learning process, which involves lived experience on the Land. Such lived experience allows understanding stories from ancestors and participating in ceremony and learning through the language (Kanien'kéha). Settlers may never get to understand the true essence of Onkwehón:we ways of understanding Mother Earth. The purpose of sharing our own process of coming to know here is to recognize the role Art plays in diminishing the effects of colonial logic frontiers in science education.

AMY'S STORY: ART AS THE APPRECIATION OF MULTIPLE WAYS OF COMING TO KNOW

I recently participated in a land-based wilderness survival skill camp led by a few teachers from Kahnawà:ke. Being in nature, being in the community–learning happened spontaneously yet so smoothly. The notion of "disciplines" did not exist. Hence, the terms "interdisciplinary" or "transdisciplinary" do not exist in the context of the teacher's pedagogy. A true format of what Ted Aoki referred to as *curriculum-as-lived* or what Herman Mitchell referred as "Land-based" or what Kanien'kehá:ka people call as "Tsi Niionkwarihò:ten curriculum". The teacher was living the curriculum. Relationship with the land, relationship with each other, relationship with the knowledge holder—the learning happens through relationships. Such relationship-based, relationship-focused curriculum encourages the "multiple ways our human sense of living together" (Donald, 2009, p. 8) consist of "mixed understandings of these places" (Donald, 2009, p. 10).

As a teacher-educator, I strive to provide similar learning opportunities that I've experienced in the camp. Learning that is contextualized in specific place/Land (land-based); facilitates relationship-building processes and encourages looking at the relationship between diverse ideas (relationship-based) and, finally where knowledge is enacted and shared by action, thus becomes wisdom (wisdom-based). This approach allows for learning with one's 3H, that appreciates multiple ways of coming to know nature in their mind (head), engages one's emotions (hearts), and allows continual sharing of ideas and lessons amongst the learning community (hands).

Reflecting on diverse pedagogy for science teacher education, I came across David Blades (2015) work in aesthetics in science education. David developed an assignment for his pre-service teachers where they explore the "beauty of nature" through artistic and scientific ways (Blades, 2015). Inspired by his work as well as my own learning in the Kahnawake community through my role in science education consultant, I've invited pre-service teachers enrolled in the course to engage in the assignment, titled, *Engaging 3H: Appreciating the beauty of nature through multiple ways of coming to know.* My purpose was to create a self-directed learning opportunity for my students to first-hand experience artful-scientific inquiry and understand that art and science both have a systematic research/inquiry process which involves creativity and collecting, processing, and analyzing and communicating ideas. In this 3H assignment, I particularly focused on the *process* of engaging 3H- head (mind), heart (emotion), & hands (action). Thus, students were encouraged to document the whole processes of their artful-scientific inquiry, including their emotional changes and engagement with WMS and Onkwehón:we ways of coming to know throughout the process. On the last day of class, we had a sharing symposium where we invited the children of our students to participate. Sat in circle, everyone shared the product and the process of their artful-scientific inquiry. In the middle of circle, children were sitting down, listening to their parents and Elders speak. Situating myself a learner, I sat in circle and listened quietly.

HANNAH'S STORY: ART AS A STARTER OF ARTFUL-SCIENTIFIC INQUIRY

Every morning since my Grandmother died in December 2016, my Grandfather comes over to my house and he, my mother, and I have breakfast together. While we are enjoying this meal, the birds, squirrels, and rabbits join us, dining on seeds and corn from the feeders and scattered across the back porch. Of all the creatures that visit our home, the cardinals, in particular the males, are the most remarkable for their vibrant colors and feisty tempers. Cardinals were also my Grandmother's favourite bird so it is always a treat when we catch a glimpse of the birds she loved so dearly. We have had this belief in my family that when cardinals visit your home, it is the spirits of loved ones coming to check in on us and let us know they are still watching. In a way, it feels like we are still sharing a meal with those we have lost when the cardinals come to call. I attempted to capture their beauty and mystery in a painting at the behest of my Grandfather just before last Christmas (see Figure 10.1).

Figure 10.1. Cardinals.

Source: Hannah Karahkwenhawe Stacey

I remember thinking while I was painting, the cardinals should not be able to survive, particularly in winter. The males, so starkly contrasting the woods and snow of their winter habitats, radiate with a flare of brightness that I would assume would be a signal of food for hungry predators. The females, with their brown and tan coloring can easily blend into the trees, but the males have no such camouflage. They can be seen from miles away perched in the trees. How, then, have they survived predation and not gone extinct or evolved to have a more muted color? How has the male cardinal survived predation?

Thinking back on what I know of similarly vividly colored creatures; lady-bugs, caterpillars, and the like, I wondered if perhaps predators avoid them thinking them poisonous. I felt silly thinking it as soon as the thought formed. Logically, it would require that the predator have a diet that included the afore-mentioned insects as well as birds as large as cardinals. While not impossible to think of a predator that could fit this category, it would require them to have partaken of plenty of the poisonous insects prior to contact with cardinals, which seems unlikely. Moreover, the predator would have to possess logical thinking and color differentiation almost on par with humans to assume that vivid color is equated with toxicity.

I then thought of butterflies who are conspicuous when viewed from one angle, but disguised when viewed from another. As I observed the male cardinals, I did notice that their wings and backs are less intensely colored than their fronts, composed of more browns and tans with the red. Still, the color difference is not enough to allow them to blend in easily, and it would require they turn their backs on potential threats. I could not help but be unsatisfied with my deductions and hypotheses and needed to research.

I want to take a step back for a second and say that I am not a stranger to research. I spent years studying science in high school, Collège d'enseignement general et professionnel (CEGEP) and University (the first time around) and then doing research papers for college. During the senior years in high school, I decided to choose to further study in the field of science. I chose it because I liked that it seemed to always have one correct answer. In high school, this is true enough. Most high school science is simple, perfect problems that can, and have been, solved many times over. As we ventured into more and more complex questions, things do not always work out that way and more than one conclusion may be right. I have learned that science and "fact finding" is a process that is always ongoing and never complete and enough. We simply pass on, as fact, what we have the largest consensus is true and has the most evidence to support it. I always try to keep this in mind, and in turn keep an open mind, when I venture to find answers to my questions.

A quick, online search of cardinals was very unhelpful to say the least. I found absolutely nothing to confirm or deny my own questionable hypotheses on why male cardinals haven't suffered extinction due to predation, nor any new ideas on the subject. I was dismayed. I genuinely thought this would be an easy topic of research, but I persisted. Finally, I found one source that related the information I sought.

On an internet page, allaboutbirds.org, hosted by Cornell University, I found someone had already asked the same question as I had, "How does the male [cardinal] get away with being so colorful?" The answer they came up with surprised me, but seeing it came from a university affiliated website, I decided to trust the research. Apparently, the red is a bigger positive in the bird world than a negative. While the color does make the male cardinal a target for predation, it is also a mark of honor among birds for reproductive purposes. The simple answer is, male cardinals breed faster and more frequently than they can be hunted. Because they do not migrate they have a longer breeding season and can have multiple births a year. The brighter red the male cardinal, the more virile the cardinal is seen to be by the females. The potential for bright color is then passed on to male offspring who deepen the shade due to a diet of berries (among other things but it's the berries that give them the color). There is no need for camouflage when you can just make more male cardinals than can be preyed upon.

A simple question that occurred to me while creating a painting for my Grandfather led me to a surprising conclusion that I had to hunt for through pages of interesting, if not what I was looking for, information on the internet. In general, when we set out to answer a question, we are not always guaranteed an answer. We can make educated guesses based on prior knowledge and observation, but those can only get us so far. We need to research and be wary of the research we find because not everything presented can be trusted. It may take a long time to wade through the "unimportant" or "irrelevant" information to find that nugget of information we seek, but the journey is worth it if the question is "important" or "inspiring" enough. I have come to see all this in the brief research I did for this paper and by reflecting on my own thought process of the experience. These are the things I would like to share with my future students; persistence, pursuit, and assessment of the information we attain is all a part of scientific-artful inquiry and it is rewarding when you are inspired by your questions–that comes from living on the Land.

RO'NIKONHKÁTSTE'S STORY: ART AS A PROCESSING AND COMMUNICATING IDEAS

<u>Tsi Niionkwá:nakte</u> – <u>Interpretive translation</u>
Our Place
Tsi niionkwá:nakte, kénh niioto'ktátie,
Our place is, up to this point,
Ionkwanaktiiò:se, tsi niiohtentionhátie.
Comfortable for us, with the way things are going.
Ken?
Is it?

Hen, ion'wé:sen tsi niionkwarenhnhà:'on
Yes, what we are used to is enjoyable
nek tsi tó:kenhske ken,
but is it true,
Tsi kanaktí:io í:we?
that it is a nice place?

Aietewatatenaktóthahse'
We best make time for ourselves
Ohén:ton iaietewanonhtónnionhwe',
To think towards the future
"Tó: nikarì:wes enionkwanaktó:take'?
"How long will we have?
Enionaktó:take ken?"
Will there even be room?"
Iah se' é:so tetsonkwanáktote.
It seems there is not much time left.

Aonsonterihwakwarihsi' tsi niionkwaié:ren',
What we have done must be made right,
Aonsakakwatákwenke tsi nenhotinaktó:take
Their place must be fixed
Ne ohén:ton taiawenhniseratátie
For the days coming forth
Oh naiá:wen'ne' ronónha
So that they too
Enhotinaktiio'sè:sheke'
Will remain in a nice place

Ka' né: nòn:we, òn:we,
Where did it go
Tsi niionkwanaktotà:'on?
The time we had?
Tóhsa thó: nienhén:we'
It should not go there
ne aonsahonatò:kthahse'
where they will have none left
Tsi nén:we'.
Forever.

As a child, I gazed up at the stars identifying constellations, curious to know what lies beyond. "How big is the universe?" A question that guided me to further investigate not only the science behind the universe' existence, but my own existence within the universe, from an Onkwehón:we perspective.

My journey began with curiosity. I wondered how large the universe truly is, and furthermore what role do we play in it? How large of a role do we play in the grand scheme of this immense concept? Through my research, I learned that in 2013 a study found that the oldest observable source of light in the universe was 13.8 billion years old. Scientists then took these findings and under the assumptions that the universe is spherical in shape, found the radius of the *observable* universe to be 28 billion light years. A light year is the measurement of distance that light can travel in the duration of one year (Redd, 2017).

Upon learning of the immensity of the observable universe, I was inspired to further investigate and attempt to comprehend how large the universe could possibly be beyond that which is observable to us. In my research, I learned that through working on Einstein's theory of gravity, in 1929 Edwin Hubble discovered that the universe has been expanding from a single point of origin, which we now understand as the big bang theory. Furthermore, I learned that in the early 20th century, upon the discovery of relativity by Einstein, it was revealed that space and time were relative to each other, as opposed to being independent (Overduin, 2007). This means that the size of the universe is, and has been for billions of years, expanding. Even more amazing, is that this expansion of space is literally time unfolding.

It became clear that our role within this unimaginably vast universe was not insignificant after all. In fact, our role is the growth, expansion, and unfolding of it. Every moment that passes, every move we make, is the universe growing. After viewing it from this perspective, a mind-blowing sensation of connection and purpose overcame me. We do not exist within the universe, but rather the universe exists within us, we *are* the universe!

My definition of science prior to this research was the understanding and exploration of the world around us, as well as the role we play within nature. When thinking about science, I often feel a sense of connection to the natural elements with which we co-exist, and curiosity overcomes me as I seek to understand and maintain the relationships we share. I chose to utilize this scientific learning experience as an opportunity to further investigate my understanding of scientific concepts from an Onkwehón:we lens. I had believed that it would be beneficial to approach science and education from an Indigenous epistemology, and thus sought to validate this approach by further examining concepts of space and time.

As I was researching, I began to once again make connections to my own language, Kanien'kéha, and the universal view that Kanien'kehá:ka (Mohawk people) hold. As a language learner, instructor, and advocate, I spend a great deal of time dissecting and attempting to understand the worldview that was held by the original speakers of my language. It appears that the underlying concepts of these discoveries were previously understood by Kanien'kehá:ka. Within the language, it becomes evident that the original speakers of it had a thorough understanding of their relationship to the universe and the role they played in the nature of time and space. Listed below are some examples that illustrate the understanding of our existence within time and space, which are one in the same.

EXAMPLE 1:

Ionkwanaktiiò:se
Interpretive understanding: 'we are comfortable'
Literal understanding: 'the space/time is nice to us'
It.to.us-space/time-nice-benefactive

EXAMPLE 2:

Enionkwanaktó:take'
Interpretive understanding: 'we will have time'
Literal understanding: 'time/space will be erected to us'
Future-it.to.us-space/time-erect-continuative

EXAMPLE 3:

Enionaktó:take
Interpretive understanding: 'there will be room'
Literal understanding: 'there will be space/time erected'

Future-it-space/time-erect-continuative

These examples demonstrate the ambiguity of the root word–*nakt*–which may represent both time and space. The interpretive understandings may easily overshadow the underlying concepts in the structure of the words. A first-language speaker might simply translate the words as "being comfortable", "having time to do something", or "having (physical) room for something". Upon further examination of the individual components of the words, it becomes evident that there are some very intriguing scientific concepts embedded.

Prior to conducting thorough research, I had known that the Kanien'kéha word *Kanákta* could interchangeably be interpreted as either time or space. Upon learning of the scientific nature of the universe's existence, I felt a sense of pride for the sheer brilliance that my ancestors possessed, as this concept was already embedded into our language. My personal definition of science remains the same, however this reaffirmed my prior observations with scientific evidence, further proving the benefits of Indigenous ideologies in education.

The implications for teaching this particular topic are not only to understand that time and space are one in the same, but to show students that our very existence is the expansion of the universe in progress. Jacobs (2003) states, "as can be understood through our language, it is the relationship to the natural world, the active participation with it, that are key in Indigenous science" (p. 48). In an Indigenous perspective, the implications may re-establish lost connections, understandings, and approaches to contemporary knowledge, all which are crucial to the survival of Indigenous epistemology in modern society.

I am once again reminded of the importance to teach children Kanien'kéha as well as Onkwehón:we way of thinking. The implications of appreciating the beauty of science, and therefore nature. This approach will ensure that connections to the natural world are maintained and respected. Concepts such as space/time, and its translation into our existence, will not be lost. Furthermore, in contemporary society I think that such an intriguing discovery may even spark interest to further investigations of what science knowledges the Indigenous people of this land held.

I chose to create a representation of what I learned through artistic literature, or poetry. As I began to wonder what exactly I should speak on, I reflected upon the class discussions, readings, and my research and personal thoughts. An image of the message I wanted to convey became clear to me. I wanted to take the concept of space and time being the same, both in science knowledge as well as Onkwehón:we knowledge. I illustrated that concept through Kanien'kéha language, while conveying a message of the urgency to address environmental issues. This urgency is something that I believe should not only be approached from an Onkwehón:we perspective, but in Onkwehón:we language as well.

Indigenous concepts of space and time through creative writing/poetry speaks to the topic of environmental issues, thus creating cross-curricular connections. These connections include: Language arts (expressive writing/poetry), Science (environmental issues, concepts of space and time), Social Studies (Indigenous worldview, concepts of space and time). It also promotes higher level thinking, which will create enduring understandings that will stay with the students longer than traditional ideologies, that merely fill students with facts. Most importantly, it demonstrates identity curriculum within students' education, as they see themselves as a part of science and active members of the universe.

FINAL SHARING WORDS

Working with multiple ways of coming to know involves living in balance and harmony. A pedagogy that celebrates balance and harmony requires building relationships and should strive for creating a *sharing place* (Kim, 2018)—where multiple ways of coming to know are encouraged and educators and learners—together are learning without hierarchy. 3500 words is not enough space to share all the inquiry processes and lessons we learned throughout the process. We hope that our stories of engaging artful-scientific inquiry inspired others to start your own journey of critical reflection and challenge the *colonial logic frontiers* present in your own life and work, striving for balance and harmony while engaging in multiple ways of coming to know.

REFERENCES

Bishop, A. (2015). Becoming an Ally: Breaking the cycle of oppression in people. Winnipeg, MB: Fernwood Publishing.

Blades, D. (2015). Recognizing the beauty: Aesthetics in teacher education. Keynote lecture presented at Science Education Research Group, Canadian Society for the Study of Education. Retrieved from http://www.sergcanada.ca/wp-ontent/uploads/2012/05/SERG-Paper-David-Blades.pdf

Derry, C. (2005). Drawings as a research tool for self-study: An embodied method of exploring memories of childhood bullying. In C. Mitchell, K., O'Reilly-Scanlon, & S. Weber (Eds.), *Just who do we think we are?: Methodologies for autobiography and self-study in education* (pp. 34–46). New York, NY: Routledge Falmer.

Donald, D. (2009). Forts, curriculum, and indigenous Métissage: Imagining decolonization of Aboriginal-Canadian relations in educational contexts. *First Nations Perspectives, 2*(1), 1–24.

Jacobs, K. L. (2003). Towards renewed balance & harmony in the natural world: An environmental responsibility protocol for Kahanawake Mohawk territory and beyond. [Master's thesis]. Royal Roads University. Theses Databases.

Kim, E. A. (2018). The relationships at play in integrating Indigenous knowledges-sciences (IK-S) in science curriculum: A case study of Saskatchewan K-12 science curriculum. [Doctoral dissertation.] McGill University. Theses Databases.

Overduin, J. (2007, November). Einsteins spacetime. Retrieved from https://einstein.stanford.edu/SPACETIME/spacetime2.html

Redd, N. T. (2017, June 7). How big is the universe? Retrieved from https://www.space.com/24073-how-big-is-the-universe.html

Snively, G., & Corsiglia, J. (2001). Discovering indigenous science: Implications for science education. *Science Education, 85*, 6–34.

Tanaka, M. T. (2016). *Learning & teaching together: Weaving Indigenous ways of knowing into education*. Vancouver: UBC Press.

Wilson, S. (2008). *Research is ceremony: Indigenous research methods*. Halifax, NS: Fernwood Publishing.

A in STE(A)M

Indigenous underrepresentation
    negative experiences   disparate achievement
          capillaries of Eurocentrism
   voices silences   red dresses dismantled   missing Joyce
        loose fibers
              becoming lost
                   racism in binaries
to decolonize   deconstruct colonial gaze   means gazing inward
          embrace and reassess tensions
              through
participatory and reflective space of embroidered connections
    that slows us down to then   see   experience

           a holographic web of connection
            revealed through art
artto navigate a deeply cross-cultural space
slow-twinned pedagogy   playful exploration
             digital storytelling
weaving animal locks that link us to the ancestors
    tobacco as bridge   plants as teachers
art is a production of vulnerability

art disrupts normative forces
art is knowledge in constant motion
art connects   animal with human
                                                art with math
                                                individual with community
cross-pollination of art and science   means   coexistence
        response-ability   I to we
shake roots   bear witness   unfold   release boundaries   reveal
        beyond the enclosure   of the garden
                                        or the fort

De/colonizing Pedagogy and Pedagogue: Science Education Through Participatory and Reflexive Videography

MARC HIGGINS
Department of Secondary Education, Faculty of Education
University of Alberta, Edmonton, AB, Canada
Grateful acknowledgment is made to the following for permission to reproduce copyrighted material:
Higgins, M. (2014). De/colonizing pedagogy and pedagogue: Science education through participatory and reflexive videography. Canadian Journal of Science, Mathematics and Technology Education, 14(2). doi: 10.1080/14926156.2014.903321

DECOLONIZING SCIENCE EDUCATION AND SCIENCE EDUCATOR: THINKING WITH BELCZEWSKI

Within Canadian science classrooms, local Indigenous ways of coming-to-knowing[1] the naturalworld continue to be underrepresented, misrepresented, misunderstood, and undervalued,whereas Western modern science (WMS) is overrepresented and misrepresented (Aikenhead, 1997, 2006b; Aikenhead & Michell, 2011; Aikenhead & Ogawa, 2007). As a result, as Aikenhead and Elliot

(2010) stated, the culture of "school science" is foreign and one to be avoided if possible for the vast majority of students (~90% of all students)[2]. Moreover, for many Indigenous students, learning school science is often characterized by negative experiences and disparate achievement rates when compared to their non-Indigenous classmates who reproduce and uphold an underrepresentation within the fields of science and technology (Barnhardt &Kawagley, 2005, 2008; Cajete, 1999; Canadian Council on Learning, 2007a; MacIvor, 1995). The initial focus on addressing these issues within science education has primarily been one of producing and implementing culturally relevant curricular resources. However, despite some success, integrating Indigenous perspectives of science into an educational system that is shaped by Eurocentrism[3] and whiteness[4] often reproduces the same problematics, albeit differently(McKinley, 2000, 2007; Sammel, 2009)[5]. In response to these problematics, there has been an emergence of science education research in recent years that focuses on both the centering of Indigenous perspectives and the decentering of Eurocentric structures that often shape science education, a twin set of processes referred to as decolonizing science education (e.g., Aikenhead 2006c; Aikenhead & Elliot, 2010; Belczewski, 2009; Chinn, 2007; Higgins, 2010).

As Menzies (2004) stated, "A commitment to truly decolonized research must be more than fine words: it must be an act and demonstrable in practice" (p. 17). However, the act of translating curricular theories to pedagogical practices is a complex and complicated process in which something is always lost in translation. The multiplicity of shifting, differential, and contextual relationships always already exceeds that which is and can be accounted for, with, and by theory. Furthermore, because Eurocentrism and whiteness circulate in capillary and often invisible manners, there is a plurality of moments in which decolonizing curriculum is exceeded by pedagogy and requires active negotiation and navigation between Indigenous and Western epistemologies and ontologies (Madden & McGregor, 2013; Nakata, 2007a, 2007b). As such, decolonizing science education must entail not only curricular reconfiguration but pedagogical practices that disrupt concepts and categories that create, and are utilized to uphold, inequality within science education, as well as the systems under which these inequalities become possible.

In order to focus on such unsettling pedagogical practices, this article draws on and extends Belczewski's (2009) study that presents a model for decolonizing science education by a white science educator through the retroactive examination of her own decolonizing journey of participating in informal science education in Aboriginal communities. Of considerable importance within this piece is her inclusion of the science educator within the scope of the decolonial gaze she was employing to analyze science education. The gaze inward permits Belczewski to think about and through the ways in which she was systemically implicated within practices of Eurocentrism and whiteness that occur in the context of science education. Furthermore, it allowed her to engage in practices that allow her to reconfigure her ways-of-being a science educator and, accordingly, the education she facilitates. She has more recently stated that this inward gaze is a necessary part of the process:

> The beauty of working with Aboriginal people is the call to self-reflection and change in our way of thinking. We are called upon (usually implicitly, but often explicitly) to think about how we think and to challenge our assumptions. (Andrea Belczewski, personal communication, April 23, 2010)

As a *Qallunaak* (i.e., non-Inuit), white teacher, teacher educator, and educational researcher doing work in Nunavut, it is productive for me to think with Belczewski's (2009) theorizing on decolonizing science education and science educator to uncover some of the ways both are implicated in the process of replicating Eurocentrism and whiteness that they work against. Herein, I revisit a science education project that worked toward decolonizing the ways-of-knowing and ways-of-being that shape both science education and myself as science educator in order to make space for Indigenous knowledges, knowledge holders, and pedagogies (Higgins, 2010, 2011). In parallel, I illuminate the partial and complex failure in translation that occurs between decolonizing theories (i.e., border crossing and reflexivity for decolonial purposes) and associated pedagogical practices of decolonizing science education.

RESEARCH CONTEXT AND DESIGN

Research Context

The research study took place in Iqaluit, the capital city of Nunavut. For youth living here,the navigation between Inuit *Qaujimajatuqangit* (IQ) and WMS is one that occurs daily. IQ here refers to "all aspects of traditional Inuit culture including values, world-view, language, social organization, knowledge, life skills, perceptions and expectations" (Nunavut Social Development Council, 1998, p. ;1). As the hub for circumpolar research in the eastern Canadian Arctic, Iqaluit is both an environment that is rich in Inuit traditional knowledge as well as Western scientific practices. It is increasingly recognized that IQ and WMS do not have to be viewed as mutually exclusive because there are many scientific concepts from both worldviews that resonate with one another(Watt-Cloutier, 2004; see also Peat, 2002). A growing body of research works toward the decolonizing project of integrating and centering IQ within science education that mimics the already occurring cross-cultural exchange in the region (Higgins, 2010, 2011;

Lewthwaite & McMillan, 2007; Lewthwaite, McMillan, Renaud, Hainnu, & McDonald, 2010; Lewthwaite & Renaud, 2009).

Twin-Lens Methodology

In order to explore how youth navigate and negotiate the space between IQ and WMS within science education as one example of decolonizing science education, I employed dual decolonizing research processes that I am referring to as a *twin-lens methodology* (Higgins, 2010). This methodology utilized participant-directed videography as pedagogy to explore and value participants' perspectives and perspectivities (Goldman, 1998; Riecken et al., 2006). Simultaneously, it uses self-reflexive video diaries, which Dowmunt (2011) described as the "'to-camera piece' in which the video diarist turns the camera on her- or himself and records her/his thoughts" (p. 171), to examine what I implicitly permitted and prohibited with respect to the space in between IQ and WMS as a pedagogue implicated with/in Eurocentrism and whiteness. Though the twin-lens methodology is described and discussed in greater length elsewhere (Higgins, 2010), the use of visual research methods in extending the dual processes of decolonizing education and educator, as well as centering Indigeneity and decentering Eurocentrism and whiteness is significant. Given that educational institutions, materials, pedagogies, and curricula are almost always "founded on a vision and visualization of education and culture that look to Europe as the centre of all knowledge and civilization" such that a "Eurocentric curriculum is hidden in plain view" (Battiste, Bell, Findlay, Findlay, & Henderson, 2005, p. 8) and that Western modern culture is occularcentric (i.e., giving vision primacy), visual methods and methodologies provide prime and productive sites from which to examine the complex systems under which these inequalities become possible.

Video Data Collection

The research was conducted during a 2-week long informal movie-making summer program focusing on science and technology at the Nunavut Arctic College in Iqaluit, which was delivered in partnership with the Nunavut Research Institute (see http://www.nri.nu.ca) and Actua (see http://www.actua.ca). All youth enrolled within the program opted to participate within the research. In this program, participants who were 12–15 years of age, both Inuit (n = 4, In) and non-Inuit (*n* = 3, nI), male (*n* = 4) and female (*n* = 3), learned the basic techniques and skills required to create their own video-media in order to explore, define, and document the ways that science is enacted in their communities. Through this participant-directed videography project, youth generated their own short movies which were a form of digital storytelling, as well as documentary-style video interviews with community participants working within science and technology related fields (n = 7; 30–60 minutes each). During their community interviews, youth participants developed and delivered questions around issues and daily uses of science and technology in Iqaluit. These interviews were paired down through video editing by participants into 5-minute vignettes. During the development of their own short movies, youth employed all the elements of preproduction (e.g., storyboarding, shot lists), production (e.g., sound, lighting, camera work), and postproduction (e.g., video editing, sound score). Concurrently, I regularly produced video diaries that were utilized to analyze myself as decolonizing pedagogue. These reflexive video diaries were produced at the end of every day during the 2-week-long moviemaking program and lasted approximately 10 minutes each. In addition, a few additional reflexive video diary sessions took place the week before and after the program. In total, approximately 150 minutes of reflexive video data was collected. Lastly, these video diaries were supplemented by reflexive journaling that usually but not always followed the reflexive video work.

Video Data Analysis

In order to make meaning of and with the video data disposal, I used rhizomatic analysis (e.g.,Gough, 2006) through mind-mapping methods (e.g., Wheeldon & Faubert, 2009). Rhizomatic analysis works against the reductionism of conventional coding methods through productively putting to work a network of connectivities and relationships between theories, practices, data, ethics, and other bodies of knowledge and being that are always already becoming. Because rhizomatic analysis can be seen as a way of methodologically *getting lost* (Lather, 2007), notas losing one's way (i.e., losing sense of where one is and where one might go) but rather as losing *the way* (i.e., losing sense of there being *a* way that is singular and definite). As such, it becomes an appropriate choice of analytical mode for a project that attempts to decenterdominant analytical modes shaped by Eurocentrism and whiteness. As a method, mind-mapping temporarily and partially represents such a web of connectivities. Using *Visual Understanding Environments*, a mind-mapping software program whose grammar consists of nodes and links, Iwove an intricate web of inter- and intrarelationships between data chunks (e.g., video, textual),concepts, and theories[6]. This method productively permitted the exploration of this generative knowledge space that is within, in between, and beyond both participatory and reflexive video data sets, as well as assisted in the identification of significant regions within this impermanentmetadata mapping. The discussion below highlights a few of these resonating regions that actedas points of indetermination or locations that exceed and resist "too easy" forms of signification(Lather, 2007). In particular, the focus is to illuminate complexities and complications that occur when pedagogical practice exceeds curricular theorizing within decolonizing science education.

Decolonizing Pedagogies: Exceeding Border Crossing Through Participatory Videography

Within the science education literature, the predominant pedagogy or pedagogical metaphor that is utilized for decolonial goals is that of *border crossing*. Aikenhead's (e.g., 1997, 2001a, 2006a,2006b) prominent use of Giroux's (1992, 2005) border crossing concept has made it synonymous with the cultural borders that learners who do not see themselves within the world of WMS must cross in order to succeed. As Aikenhead (2001b) stated, "Success at learning the knowledge of nature of another culture depends, in part, on how smoothly one crosses cultural borders" (p. 340). Furthermore, for both Indigenous and non-Indigenous youth to learn to negotiate and navigate the crossing between worldviews is to develop a high level of flexibility in thought that is an asset in today's society.

As a cautionary note, it is important to consider that border crossing-based learning should never be at the expense of the learners' own personal or cultural ways-of-knowing or ways-of being (Aikenhead, 2001a; Barnhardt & Kawagley, 2008). Furthermore, Sammel (2009) cautions greatly that this type of educational metaphor and the practices that stem from it will continue to uphold and reproduce Eurocentrism and whiteness if their final objective is a unidirectional border crossing of youth into the *culture of power* because of the assimilative nature of the act. Through educational enculturation, students are often implicitly told what they can gain through unidirectional border crossing but seldom do educators ask what can be gained by crossing bi- or multi-directionally. There is much that we can all learn from one another when we engage in multidirectional border crossing (Barnhardt & Kawagley, 2005, 2008; Peat, 2002). However, if we are to use border crossing as a metaphor, it is important to heed Grande's (2004) statement that border crossing, for many Indigenous students, never is a choice. They "did not 'choose' to ignore, resist, transcend, and/or transgress the borders" (p. 167); rather, they were placed there when the imaginary lines of power were drawn in the sand. An effective and appropriate use of border crossing as educational metaphor must come with a critical analysis of the borders themselves: how they were placed there, what systems of power they uphold, and how crossing them is framed.

Given that participatory videography centers an individual's complex, complementary, and conflicting points-of-viewing the world, it is a rich method for exploring how participants negotiate and navigate the deeply cross-cultural space in between IQ and WMS in their community of Iqaluit, Nunavut. Points-of-viewing, as Goldman (1998) stated, "encompasses where we are located in time, space, as well as how our combination of gender identities, classes, races, and cultures situates our understanding of what we see and what we validate" (pp. 3–4). Because points-of-viewing include these differently situated cultural locations, it provides a rich site of engagement from which to decolonize pedagogies for science education by using and troubling border crossing. Here, youths' points-of-viewing science are explored through two different types of videography endeavors: interviews with community members and video-based storytelling. I could not speak about the pedagogy of participatory video without mentioning the navigation and negotiation of how the project was introduced. When I informed the youth that their primary task was to create videos that explicitly center science, I was met with a silent series of scrunched brows (i.e., how "no" is bodily conveyed in Inuit culture). I could not have imagined that participants' primary perceptions of science were equated with a school science that is "probably for nerds" or a stereotypical image of scientist that effuses from contemporary media representations (i.e., lab-coat-and-goggle-wearing, middle-aged white males whose time is spent in the laboratory) and not the rich cross-cultural examples I noted in Iqaluit (Higgins, 2011). At that moment, a difficult pedagogical choice presented itself. Whereas decolonizing pedagogies often present themselves as what is best systemically (e.g., disrupting Eurocentrism and whiteness through a science education model based on Indigenous knowledge systems) as well as individually (e.g., empowerment through counterstories of how students cross borders with respect to science in their daily lives), I encountered a situation in which the research would be upholding Eurocentrism and whiteness one way or the other. In other words, were I to choose to continue with the video project I designed, though it would potentially generate systemic changes, it would be a form of pedagogical violence through the imposition of a colonizing teacher–student relationship in which knowledge flows unidirectionally (Kanu, 2011; Rasmussen, 2002). On the other hand, were I to drop the project, it would respect individual participants but allow the status quo shaped by Eurocentrism and whiteness (i.e., conceptions of "what counts as science") to remain unchallenged. When the systemic and the individual do not align in decolonizing pedagogies, the binary between decolonizing and colonizing begins to break down. In other words, decolonizing becomes in part colonizing. In this situation, the participants and I negotiated something that was in between rather than one or the other. This resulted in participants agreeing to participate in already scheduled interviews with local Inuit and non-Inuit professionals working within the field of science and technology, while I agreed that the media-based storytelling would be free-form in terms of content.

Through the community interviews, youth went "on location" to the Inuit Broadcasting Corporation ($n = 1$), the Canadian Broadcasting Corporation ($n = 3$), the Nunavut Research Institute ($n = 2$), and the hospital ($n = 1$) to engage both Inuit (In) and non-Inuit (nI) professionals in discussions focused on how they navigated and negotiated the space between IQ and WMS within their science and/or technology related profession. In preparation for the interviews, participants developed questions (e.g., "What are some of the steps required to attain the job you have?") in order to direct the conversation in directions that they deemed productive. Arriving upon location, the participants' excitement was palpable. Notably, one of the participants cried out "expert!" (i.e., a local expression that might be analogous to "awesome" or "cool" while carrying the significance of expertise, professionalism, and mastery) during a visit to the Inuit Broadcasting Company, because she was excited about accessing the people and place where local media was being produced. These interviews provided insight into how both Inuit and non-Inuit science and technology professionals negotiate the space between IQ and WMS. Notable examples of this include the Nunavut head nurse's (nI) realization of how Inuit women often negotiate pain differently during birthing than many non-Inuit women (i.e., by remaining strong and silent). In addition, how videographers (In) at both the Inuit Broadcasting Corporation and Canadian Broadcasting Corporation seamlessly infuse IQ into the process of videography, relationships inherent to media-making (i.e., utilizing Inuit protocols in relationship-building), as well as in the content (i.e., Inuit storytelling, traditional foods cooking show). Lastly, how scientists (nI) at the Nunavut Research Institute work toward honoring and putting to work local knowledges in developing ecological and geological understandings of the Arctic (e.g., by engaging with Elders and traditional storytelling to map out ecological systems). Not only were these examples notable through the interviewing process but they also made the final cut through the editing process. They were utilized to paint a more complex portrait of the cultural locations of various Inuit and non-Inuit professionals working within fields that are typically or usually represented within the fields of science, media, and technology (e.g., scientist, nurse, videographer). Participants presented professional bodies that did not sit squarely within cultural categories of Indigenous or non-Indigenous but rather were positioned somewhere in between. Furthermore, learning from community members working in science and technology-related professions contributed to an understanding of how scientific knowledges and technologies are not neutral but rather are (re)shaped by the values of those using them (Harding, 2008; Nakata, 2007a, 2007b). Though participants primarily thought of science as what Pomeroy (1994) referred to as *white male science*, both Inuit and non-Inuit participants were operating with/in cross-cultural conceptions of science that spoke of science as embedded within everyday life such that anyone and everyone could engage with/in science (Higgins, 2011). As such, participants demonstrated their points-of viewing science through highlighting and centering conceptions that were congruent with their own or that stood out for them, while cutting out those that they deemed less significant through this process (e.g., technical details of professions). This is especially significant because less than 10% of raw footage made it into the final cut. With these videos, participants were able to paint a broader and more nuanced portrait of "what counts as science" within their community of Iqaluit, Nunavut, that also demonstrated how the points of resonance and divergence between IQ and WMS differ in significant ways between health care services, techno-media, and ecological research.

Though this project was framed utilizing Aikenhead's (e.g., 1997, 2001a, 2006a, 2006b) border crossing as decolonizing pedagogical metaphor, it became evident that it would be problematic to employ this metaphor in tandem with binary thinking. As evidenced from interviewees as well as the participants' points-of-viewing science demonstrated through editing, neither sits fully within either an Indigenous or Western worldview. Accordingly, to impose a system of categorization in which participants, as well as their interviewees, were either Inuit or non-Inuit would be a pedagogical form of colonial violence that would silence the ways in which cultural lines have been reconfigured within their community as well as the ways in which culture is always already in relational flux (Cajete, 1999).

Though this pedagogy runs colonizing risks, it also offers additional decolonizing possibilities above and beyond making space for youth to work within, against, and beyond the dominant and normalizing discourse of school science. For example, this pedagogy disrupts dominant linear and hierarchal pedagogies by beginning a movement toward community as pedagogy (Kanu, 2011). As Kanu (2011) stated, "Integration [of Indigenous perspectives] will be implemented in classrooms most richly if ... . [it] can be made to include [Indigenous] peoples in positive and empowering ways" (p. 115). Inviting, including, and honoring Indigenous community members who bring with them living knowledge within pedagogical processes reconfigures the way a community of learning is built (Aikenhead, 2006b; Cajete, 1999; Kanu, 2011). Furthermore, leaving the science classroom or laboratory to seek various knowledge holders within the community disrupts Eurocentric notions of schools as the centered space of knowledge as well as those of knowledge as static and fixed (see Rasmussen, 2002). This is in part because video is "able to provide the means of giving the participants an authoritative: voice" (Bloustien & Baker, 2003, p. 72), thus dampening the singularized authority of formal places of learning.

Figure 11.1. Screenshot from Participants' *Video Legends*.
Source: Author

Though I had initially written off the participants' free-form media-based storytelling asscience education (e.g., one movie imagined Oreo eating as an Olympic sport), it is importantto note that though the videos produced were not explicitly centering science, this does notnecessarily exclude its presence from within the content. One notable video-based story entitled*Legends* (re)tells three traditional Inuit stories in a contemporary context through the medium ofmodern horror movies (e.g., movies such as *The Grudge and Saw* that employ suspenseful quick-cuts that withhold information until the last minute and utilize dark and dramatic lighting andmusic). The first story (re)told the story of Muhaha, the aptly named traditional Inuit bedtime-story monster who chases after children to tickle them to death with his long claws (see Figure 11.1). Within this story, one of the lessons that can be learned is that the land (i.e., where Muhaha captures his prey) can be a dangerous place in the eastern Arctic. As such, we find here the persistent Indigenous teaching that knowledge about and knowledge practices with the ecology of place are and have always been a question of survival (Barnhardt & Kawagley, 2005, 2008; Cajete, 1999). The second story (re)told is that of a brother and sister returning from a fishing trip who meet their demise at the hands (or paws) of a shape-shifter. Here, a viewer might pay attention to locally developed ways-of-living with nature explicitly presented throughout this story such as fishing, as well implicitly presented such as Inuksuk building (i.e., the story was (re)told around an Inuksuk). In addition, the inherent knowledge of the local ecology is presented through the various shapes (e.g., polar bear, wolf, raven, human) that the shape-shifter takes during the story. The third and last story (re)tells the tale of a girl who played string games (e.g., cat's cradle) to excess. Not heeding her mother's warning, she inadvertently invited Tutaarak, the spirit of the string, through this excessive play and eventually lost her life in a death match. This game demonstrates intricate embodied mathematics present within the string games played. Furthermore, one might take note of the importance of balance and the danger of hubris and excess, all of which have been Indigenous critiques of science (Barnhardt & Kawagley, 2005, 2008; Cajete, 1999). What one might pay attention to or not is, however, up to the viewer. This is the pedagogical potential and possibility that Indigenous storytelling offers. It is a learner-centered pedagogy in which meanings are not ready made: different viewers might walk away with different meanings and teachings. Furthermore, as Barnhardt and Kawagley (2008) remind us, traditional stories are rich sites of knowledge about the natural world (i.e., what is usually referred to as science). However, this knowledge cannot be fragmented from the whole as "the native creative mythology deals with the whole physical, intellectual, emotional and spiritual aspects of these inner and outer ecologies" (p. 231). Furthermore, to fragment such knowledge risks perceiving these stories "only through the categories Western science has imposed on them," categories which, through Eurocentrism operate to "impose limits on what counts as knowledge and what does not" (Battiste, 2005, p. 131).

Though this video works toward decolonizing goals (e.g., resisting Indigenous/Western and modern/traditional binaries, resisting the fragmentation of holistic knowledge), this video storytelling is also significant in rethinking decolonizing pedagogies as well as the border crossing metaphor that is often utilized in tandem. As Sammel (2009) reminds us, even culturally responsive forms of science education continue to operate under in assimilative ways by giving primacy to bringing learners into "Western science ontology" (p. 653). Indigenous storytelling not only resists this particular destination but also resists modes of learning in which there is a (i.e., singular) destination for learners. Of science education it asks not "How can we get students there in more culturally appropriate ways?" but, rather, "What would it mean to think of learning as a relational journey rather than a destination?"

Decolonizing Pedagogue: Exceeding Reflexivity By Turning the Lens Inward on Whiteness and Eurocentrism Through Reflexive Video Diaries

When Eurocentrism and whiteness shape the way in which culturally relevant science education resources are utilized, the science that is delivered remains deeply problematic (McKinley, 2007; Sammel, 2009). Eurocentrism and whiteness act as barriers to meaningful engagement with Indigenous ways of coming-to-knowing by shaping how teachers interact with them (e.g., fitting Indigenous knowledges and pedagogies within Western frameworks). As Sammel (2009) discussed, Eurocentrism and whiteness "insidiously [seek] to maintain the status quo" (p. 651) by (re) structuring and (re)directing the multiple ways in which dominance is maintained. For example, the oft-included mandate of improving scientific literacy remains the same even when Indigenous knowledges are included such that the implicit goal of science education remains assimilation. As such, WMS is continuously (re)produced as the meter stick against which all other knowledges are measured (Lewis & Aikenhead, 2001).

It is also worth noting that education is not only shaped by Eurocentrism and whiteness but that educational spaces are populated by the bodies that it privileges: Across Canadian schools, the majority of teachers (i.e., 90% or more) are non-Indigenous and white (Kanu, 2011). Within science education, white teachers have been primarily characterized as holding a deficit view of Indigenous knowledge, not recognizing WMS as cultural knowledge, and underappreciating the ways in which culture and learning are intrinsically linked (Aikenhead & Huntley, 1999).

What needs disrupting are not only the concepts and categories that create, and are utilized to uphold, inequality within science education but also the systems under which these inequalities become possible. Specifically, there is a need for an examination of how Eurocentrism and whiteness shape the educational system, curricula, as well as the many white bodies that occupy and uphold it within science education (McKinley, 2000, 2005, 2007; Sammel, 2009)[7].

For those participating in, (re)producing, and (re)produced by the culture of power, the oft cited "solution" is to engage within the process of reflexivity. Here, this would entail white, non-Indigenous scholars (such as myself) working in Indigenous contexts or with Indigenous knowledges to be critical of how they construct of the world within the context of science education in order to avoid reproducing problematic relations shaped by Eurocentrism and whiteness. However, this is not a simple or unproblematic task. Rather, such reflexivity is an impossibility. As Patti Lather (2007) asks, how do you think about how you think without using the very thing with which you think? In addition, the mirror metaphor that is operationalized through reflexivity problematically (re)produces sameness through circularity, thus making patterns of difference imperceptible (Pillow, 2003). For example, Webb (2001) discussed how teachers engaging in reflexivity to work against racism within pedagogical practices often end up upholding racism. Though this may be an impossible and problematic task, I, and others undertaking similar work, are certainly not "off the hook" from working through interactions that occur between WMS and Indigenous ways-of-knowing, such as IQ, and Indigenous–non-Indigenous relationships. Furthermore, there is a need to differently configure reflexivity so that it "provides a chance for 'surveillance' or a monitoring of the self" (Bloustien & Baker, 2003, p. 70), which also works within and against the circularity of traditional modes of reflexivity.

With the aims of addressing and understanding the ways in which I am produced by and (re)produce whiteness and Eurocentrism, I asked myself difficult questions during my self-reflexive work. For example, How does my Western training in the world of science (i.e., in physics) differentially produce my conceptions of the nature of science, what it is, what it is perceived as, and what it can be? How do I work against the problematic foreclosure of such knowledge in order to maintain pedagogical flexibility? How do I work within and against the implicit Eurocentric notions of validity, empirical worth, and instrumentality that I have received in order to make space for Indigenous knowledges? Though the aforementioned questions are often unsettling, I am more so discomforted by the always present possibility of (re)producing problematic relations of power shaped by Eurocentrism and whiteness in the name of critical cross-cultural scientific literacy. Though the textualization of the video diary excerpt below fails to fully capture the audiovisual complexities within the linguistic silences (i.e., the multiple pauses [.....]), it is nonetheless a rich site to think with about the complications and complexities emerging through the reflexive work that is entailed in decolonizing science education.

> Science education definitely, in certain senses, needs to draw from … …. ideas [of creativity] … …. that it shouldn't necessarily plays within the old rules because the old rules have been a great disservice to Indigenous peoples world-wide … …. and so, you know … …. You have to start from where … …. how they [Indigenous peoples] construct their world, their ontology, their … …. their ways of knowing and their ways of learning. Which definitely … …. so I don't know. … …. While I'm not entirely sure that I, myself, would be able to tap into a kind of awareness that would go beyond cultural bounds … …. but … …. maybe … …. maybe that's what I can do as an educator … …. push … …. help push people to cross those cultural boundaries in order to raise their own awareness and knowledge … …. and maybe re-instill a bit of new creativity into the world of science. [Laughter] Which in itself seems like an [enormous] task in itself. Cause who am I to come into a community and say like … …. I don't know. … …. Here's this great thing you can gain … …. could stand to gain from using Western science as well as your own science. I'm not sure … …. like … …. there's … …. there's enough Aboriginal scholars saying that … …. that … …. that Indigenous youth stand to gain from the combination of the modern … …. "modern" Western science … …. but with the clause that they don't leave behind their own cultural science beliefs and etc. because for many Indigenous peoples, science is something that produces unfeeling aliens who … …. who in the wake of … …. of delivering or, of doing science create a great deal of human and environmental grief … …. and so. … . it's why a lot of them perceive it as negative … …. it's such a dissonance with their beliefs.

Within the above excerpt, I can see many of the themes that are largely discussed within the decolonizing science literature. For example, I find here the centering of Indigenous knowledges: "you have to start from … …. [Indigenous peoples'] ontology, their ways … …. ways of knowing and their ways of learning" and the decentering of Eurocentric knowledges: "for many Indigenous peoples, science is something that produces unfeeling aliens … …. [and is in] dissonance with their beliefs." Furthermore, there is also border crossing into the culture of power: "[Indigenous learners] could stand to gain from using Western science as well as [their] own science." However, while recognizing the need to "instill a bit of … …. creativity into the world of science" in order move toward pedagogical modes which respectfully balance IQ and WMS, I continue working within the "old rules [which] have been a great disservice to Indigenous peoples worldwide." In other words, even as I attempt to work against whiteness and Eurocentrism, I largely continue thinking within those frames. Of particular importance within this excerpt is the presence of Cartesianism, a thread that prominently runs through both Eurocentrism and whiteness. In short, Cartesianism is both the belief that various meanings and materialities are discrete quantities (e.g., mind/body) as well as the process through which they are separated from that which co-constitutes them. Through Cartesianism, Eurocentric binary modes of thought that problematically permeate science education and society at large are reproduced. Notably above is the Western/Indigenous binary that (re)produces essentializing Indigenous and Western positions, which obscures ways in which Indigenous peoples participate, not always unproblematically, within the culture of power as well as Western ways-of-being in the world that don't "create a great deal of human and environmental grief." Furthermore, it obfuscates the flux, fluidity, and relationality between and within Indigenous ways-of-knowing such as IQ and WMS as these bodies have been in a dialogical exchange for years, albeit in complex and complicated ways. In short, I am reifying that which I am explicitly working against (i.e., Eurocentrism and whiteness), in a new (video) context/medium, through the process of challenging these problematic notions. As Battiste (2005) stated, "Eurocentrism is not like a prejudice from which informed peoples can elevate themselves" (p. 122).

Nonetheless, though confession is never a cure for such reification (see Pillow, 2003), it is worth noting that video dairies reconfigure the space of reflexivity in a way that offers interesting possibilities. In engaging in reflexive practices through both traditional journaling and video diaries, one of the more significant differences is that, though also present within my reflexive journaling, the disquiet in responding to the aforementioned unsettling questions was amplified within the video diaries. In the latter, my own unease in answering these questions was largely (in)audible. I would stutter and stammer on multiple occasions and the tone of my voice was deeply flattened (i.e., devoid of intonation). Most notably, entire sentences remained unfinished as I trailed off and sat silently for periods that, though usually short, occasionally stretched up to minutes. During these silences, my discomfort became all the more visible through the direct feedback loop my laptop was providing. It was common instance for my brow to be furrowed, for me to glance away from the camera from being unable to stand my own gaze, for my nostrils to flare, or for me to bite my own lip (see Figure 11.2).

Figure 11.2. Screenshot from Reflexive Video Diaries.
Source: Author

Though the textuality of the transcripts from the video diaries fails to fully capture the complexities of the embodied silences, it is worth noting that what was uttered is significantly differentially produced than the reflexive journaling in which similar questioning occurred. One of the ways in which meaning was made with these patterns of difference was through tag clouds. Steinbock, Pea, and Reeves (2007) defined *tag clouds* as "any list of words visually weighted by their relative frequencies in a source text" (p. 2). This allows for a metacomparison of what was uttered through both mediums: reflexive journaling (see Figure 11.3) and video diaries (see Figure 11.4).

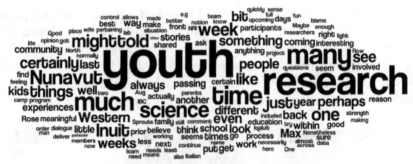

Figure 11.3. Tag Cloud of Reflexive Journaling.

Source: Author

Figure 11.4. Tag Cloud of Reflexive Video Diaries.

Source: Author

Where the tag cloud for the reflexive journaling prominently and predominantly features words one might expect in a participatory science education project in the Canadian arctic (e.g., youth, research, Nunavut, science), it is not entirely the case for the tag cloud of the video diaries. Above and beyond the predominant placement of science, the tag cloud of the video diaries had some of the following words notably displayed: *like, guess, [not] necessarily, maybe, [don't] know, kind [of], [I] think [that]*. When compared and contrasted, it can be inferred that the words I was utilizing before the camera were far more tentative as a result of the accentuated discomfort.

This is significant as the medium of video enhances "the production of vulnerability" (Dowmunt, 2011, p. 183), which is generative of the messy and irrational process of becoming other to oneself required in critical self-reflection (Lather, 2007). The production of vulnerability is enhanced in two important and interconnected ways. First, as Dowmunt (2011) explained, such a differentiation from written journaling is produced in part by the "other" who is addressed:

It felt very safe, addressing an "other" who was visibly myself. ... . In a written diary the reader, the intimate "other", has to be conjured up in the imagination, whereas in the video diary, he/she is yourself, hovering in front of you on the LCD screen. (pp. 181–182)

The reader or viewer as other whose image we must usually conjure in writing practices is not imagined but rather is replaced with the immediacy of our own image on screen when we engage in the act of giving an account of ourselves through video diaries. It is a space in which my own discomfort was no longer abstracted but rather visible before me. Second, the temporality of this reflexive process is far less drawn out than its written counterpart. When both the differential audience and temporality are considered in tandem, no longer is the reflexive process a lengthy one in which there is time to screen our reflection for academic audiences. Rather, the self as primary audience within an immediate feedback loop works against the sanitization of thought. In tandem, the discomfort produces knowledge about self differently through the disruption of normative forces that problematically (re)produce cultural myths of "good teacher" and "teacher as expert." Though still disciplined, it is a space that is differently disciplined that might productively disrupt hegemonic ways-of-seeing and ways-of-being in the world. Such a side-long glance at complex and contradictory accounts of self rather than penetrating gazes might be important when conceptualizing ways of thinking without using the very thing with which you think, even if such a thing is never (fully) possible.

Considering the Cultural Interface as a Site of (Re)Conceptualization

Decolonizing science education entails the twin processes of centering of Indigenous knowledge systems and the decentering structures and modes of thought produced through Eurocentrism and whiteness that shape science education. Working toward disrupting the concepts that (re)produce inequalities within the science classroom also entails disrupting the very systems under which these inequalities become possible. As discussed within this article, attempting to work within, against, and beyond Eurocentrism and whiteness within science education can take many shapes. It could be a pedagogical process of putting border crossing to work, critical reflexive self-work, or something entirely different. However, it is also important to apply the decolonial gaze onto the ways in which we, as science educators and educational researchers, engage in decolonizing processes. Herein, though attempts to decolonize both pedagogies and pedagogue were somewhat productive, they failed to fully account for the ways in which Eurocentrism and whiteness (re)produce cultural binaries, the systemic forms of racism produced by these binaries, as well as the ways in which I was implicated.

If we take seriously the notion that we are always already marinating in Eurocentrism (Battiste, 2005), this partial failure and loss in translation between decolonizing curricular theory and pedagogical practice is illuminated. Every attempt to work against Eurocentrism is also within Eurocentrism (i.e., to which there is no "outside") and therefore inevitably reifies Eurocentric constructs, concepts, or structures through the process. The same argument could be made for whiteness. As such, like Sammel (2009), I am beginning to "wonder if there is a science curricula that does not indoctrinate, ... if there is really an authentic 'decolonizing science practice'" (p. 653). It might be time to begin thinking of decolonization as de/colonization. However, this need not be a space of deficit (i.e., decolonizing science education as an impossibility) that requires rescue through the heroism of better and improved methods. Decolonizing curricular theories can and will never fully account for the multiplicity of shifting, differential, and contextual relationships that exceed it within pedagogical practice. Rather, though far less heroic, it can be positioned that can be deeply productive in reconfiguring decolonizing science education (e.g., How do we make science education less assimilative if the working assumption is that it is always already assimilative?). If everything is potentially problematic, it encourages critical and creative experimentation within, against, and beyond current conceptions of what currently counts as science education. For example, if we consider that the relationship between Indigenous knowledge systems and WMS is a complex relational web that is in flux as neither is immune to the influence of the other knowledge system (Cajete, 1999; Harding, 2008; McKinley, 2007)[8], we might also want to utilize metaphors that might reflect such complexity. What would it mean to consider this space in between not as a border that is to be crossed but rather as a location in which we are always already located? It may be productive to turn to Nakata (2007a), who defined such a "cultural interface" as: "a multi-layered and multi-dimensional space of dynamic relations constituted by the intersections of time, place, distance, different systems of thought, competing and contesting discourses within and between different knowledge traditions, and different systems of social, economic and political organization" (p. 199; see also Nakata, 2007b; McGloin, 2009). Though there are multiple markers of identity (e.g., gender, class, race, ethnicity) and other factors that shape (*all* of) our complex and often contradictory perspectives onto and about the natural world, our differently situated cultural interfaces could provide a rich point of engagement for decolonizing science pedagogies and pedagogues. In conclusion, it bears repeating: decolonizing science pedagogies and pedagogues is a process through which not only the concepts and categories that create, and are utilized to uphold, inequality within science education are disrupted but also the systems under which these inequalities become possible in which we are always already implicated.

NOTES

1   Within many Indigenous languages, learning is not expressed as a product but rather as a process. Accordingly, instead of speaking of Indigenous knowledge (noun), I will use Indigenous coming-to-knowing (verb). As Peat (2002) defined, "coming-to-knowing means entering into relationship with the spirit of knowledge, with plants and animals, with beings that animate dreams and visions, and with the spirit of the people" (p. 65). Coming-to-knowing and coming-to-being are ongoing and interconnected epistemological and ontological processes that are deeply relational and holistically interwoven into the fabric of everyday life (Aikenhead & Michell, 2011; Cajete, 1999; Ermine, 1998; Peat, 2002). With this deep interconnection in mind, it might be more appropriate to speak of Indigenous ways-of-living in nature within in the context of science education (Aikenhead & Michell, 2011).

2   As Aikenhead and Elliot (2010) stated, "Most students (about 90%) tend to experience school science (Grades 6–12) as a foreign culture to varying degrees, but their teachers do not treat it that way" (p. 323). See their work for more on the various qualitative and quantitative science education studies that inform this statement.

3   Lewis and Aikenhead (2001) defined Eurocentrism as "the idea that the people, places, and events of Western European cultures are superior and a standard against which other cultures should be judged." (p. 53). It is not only the "colonizer's model of the world" (Blaut, 1993, p. 10) but also a colonizing model of the world. Operating through diffusionism, a forced spread of culture, it erases or assimilates non-Eurocentric knowledge systems, establishing "the dominant group's knowledge, experience, culture and knowledge as the universal norm" (Battiste, 2005, p. 124).

4   Whiteness, as Frankenberg (1993) described it, "is a set of locations that are historically, socially, politically, and culturally produced and, moreover, are intrinsically linked to unfolding relations of dominance" (p. 6). Whiteness is a position from which white people locate themselves and others that maintains structural forms of race privilege. Through whiteness, the other is viewed through a deficit lens and white people do not see themselves as raced and cultured, rendering the structural and day-to-day racism that it produces invisible to its producers. In other words, "whiteness signals the production and reproduction of dominance rather than subordination, normativity rather than marginality, and privilege rather than disadvantage" (Frankenberg, 1993, p. 236).

5   Elsewhere (Higgins, Madden, & Korteweg, 2013), I discuss how teachers' inclusion of Indigenous content perspectives within school curriculum can be problematic without a concurrent critical examination of the systemic. For example, one teacher's attempt to incorporate Indigeneity within a science education project around sustainability resulted in the addition of "an Aboriginal perspective," a singular and monolithic vision of Indigenous peoples that is rooted in stereotypical and fixed images such as "steward of the land."

6   In other words, as multidata relational sorting software, Visual Understanding Environments allowed me to import and insert short video data clips, scholarly quotes, concepts, my own thoughts, and other textual, audio, and visual elements from the whole research project as nodes that could be visually rearranged on a canvas of indeterminate size. Furthermore, it allowed me to draw lines between them in order to map out the relationships between the various nodes within, between, and beyond the data sets in an ongoing manner (i.e., exported

mind maps were permanent but the mapping within the software could always be changed by adding/removing nodes and adding/removing links). For more information about Visual Understanding Environments, go to http://vue.tccs.tufts.edu/.

7   The statements throughout this section primarily speak to and about white, non-Indigenous science educators because this section is about (a) turning the decolonial gaze onto myself as pedagogue; and (b) addressing the similarly positioned primarily white, non-Indigenous bodies that occupy the position of science educator in Canada.

This is not to say that non-white, non-Indigenous, as well as Indigenous science educators are not implicated with/in Eurocentrism and whiteness but rather that they are positioned differently within them. Though an important site of inquiry, the multiple and distinct ways in which different bodies participate in and are affected by racism and colonization are beyond the scope of this article.

8   Castellano (2000) outlined three foundations of Indigenous knowledge: traditional knowledge (intergenerational), empirical knowledge (gained through careful observation), and revealed knowledge (acquired through dreams, visions, and intuitions). To these three Cajete (2009) added a fourth: modern knowledge. This last foundation involves the participation in "modern" practices so that it may complement the other foundations. Harding (2008) and Nakata (2007a) reminded us that it is important to recall that modernity does not always entail Western modernity. As technological products and processes are taken up, the often-held assumption is that the technology remains unaffected. Instead, what traditions these technologies suture over and, more important, how these technologies are shaped by these traditions should be considered.

REFERENCES

Aikenhead, G. S. (1997). Toward a First Nations cross-cultural science and technology curriculum. *Science Education, 81*, 217–238.

Aikenhead, G. S. (2001a). Students' ease in crossing cultural borders into school science. *Science Education, 85*, 180–188.

Aikenhead, G. S. (2001b). Integrating Western and Aboriginal sciences: Cross-cultural science teaching. *Research in Science Education, 31*, 337–355.

Aikenhead, G. S. (2006a). Cross-cultural science teaching: Rekindling traditions for Aboriginal students. In Y. Kanu (Ed.), *Curriculum as cultural practice: Postcolonial imaginations* (pp. 223–248). Toronto, ON, Canada: University of Toronto Press.

Aikenhead, G. S. (2006b). *Science education for everyday life.* London, ON, Canada: Althouse Press.

Aikenhead, G. S. (2006c). Towards decolonizing the pan-Canadian science framework. Canadian Journal of Science, *Mathematics and Technology Education, 6*(4), 387–399.

Aikenhead, G. S., & Elliot, D. (2010). An emerging decolonizing science education in Canada. *Canadian Journal of Science, Mathematics and Technology Education, 10*(4), 321–338.

Aikenhead, G. S., & Huntley, B. (1999). Teachers' views on Aboriginal students learning Western and Aboriginal science. *Canadian Journal of Native Education, 23*(2), 159–175.

Aikenhead, G. S., & Michell, H. (2011). *Bridging cultures: Indigenous and scientific ways of knowing nature.* Toronto, ON, Canada: Pearson Canada.

Aikenhead, G. S., & Ogawa, M. (2007). Indigenous knowledge and science revisited. *Cultural Studies of Science Education, 2*(3), 539–591.

Barnhardt, R.,& Kawagley, A. (2005). Indigenous knowledge systems and Alaska Native ways of knowing. *Anthropology and Education Quarterly, 36*(1), 8–23.

Barnhardt, R., & Kawagley, A. (2008). Indigenous knowledge systems and education. *Yearbook of the National Society for the Study of Education, 107*(1), 223–241.

Battiste, M. (2005). You can't be the global doctor if you're the colonial disease. In P. Tripp & L. J. Muzzin (Eds.), *Teaching as Activism* (pp. 121–133). Montreal, QC, Canada: Queen's University Press.

Battiste, M., Bell, L., Findlay, I., Findlay, L., & Henderson, J. (2005). Thinking place: Animating the Indigenous humanities in education. *The Australian Journal of Indigenous Education, 34,* 7–18.

Belczewski, A. (2009). Decolonizing science education and the science teacher: A white teacher's perspective. *Canadian Journal of Science, Mathematics and Technology Education, 9*(3), 191–202.

Blaut, J. (1993). *The colonizer's model of the world: Geographical diffusionism and Eurocentric history.* New York, NY: Guilford Press.

Bloustien, G., & Baker, S. (2003). On not talking to strangers: Researching the micro worlds of girls through visual auto-ethnographic practices. *Social Analysis, 47*(3), 64–80.

Cajete, G. A. (1999). *Igniting the sparkle: An indigenous science education model.* Durango, CO: Kivaki Press.

Cajete, G. (2009, May). *Re-building sustainable Indigenous communities.* Keynote presentation at Dream Catching conference, Winnipeg, MB, Canada.

Canadian Council on Learning. (2007a). *The cultural divide in science education for Aboriginal learners.* Retrieved from http://www.ccl-cca.ca/pdfs/LessonsInLearning/Feb-01-07-The-cultural-divide-in-science.pdf

Castellano, M. (2000). Updating Aboriginal traditions of knowledge. In G. J. Dei, B. L. Hall, & D. Goldin Rosenberg (Eds.), *Indigenous knowledges in global contexts: Multiple readings of our world* (pp. 21–36). Toronto, ON, Canada: University of Toronto Press.

Chinn, P. (2007). Decolonizing methodologies and Indigenous knowledge: The role of culture, place and personal experience in professional development. *Journal of Research in Science Teaching, 44*(9), 1247–1268.

Dowmunt, T. (2011). First impressions/video confession. *Life Writing, 2*(2), 169–185.

Ermine, W. (1998). Pedagogy from the ethos: An interview with Elder Ermine on language. In L. A. Stiffarm (Ed.), *As we see .... Aboriginal pedagogy* (pp. 9–28). Saskatoon, SK, Canada: University of Saskatchewan Extension Press.

Frankenberg, R. (1993). *White women, race matters: The social construction of whiteness.* Minneapolis, MN: University of Minnesota Press.

Giroux, H. (1992). *Border crossings: Cultural workers and the politics of education.* New York, NY: Routledge.

Giroux, H. (2005). *Border crossings: Cultural workers and the politics of education* (2nd ed.). New York, NY: Routledge.

Goldman, R. (1998). *Points of viewing: Children's thinking.* Mahwah, NJ: Lawrence Erlbaum Associates.

Gough, N. (2006). Shaking the tree, making a rhizome: Towards a nomadic geophilosophy of science education. *Educational Philosophy and Theory, 38*(5), 625–645.

Grande, S. (2004). *Red pedagogy: Native American social and political thought.* New York, NY: Rowman & Littlefield Publishers.

Harding, S. (2008). *Sciences from below: Feminisms, postcolonialities, and modernities*. Durham, NC: Duke University Press.

Higgins, M. (2010). *Decolonizing actions that speak louder than words: Science education through multiple lenses in Nunavut* (MEd thesis). Thunder Bay, ON, Canada: Lakehead University.

Higgins, M. (2011). Finding points of resonance: Nunavut students' perceptions of science. *Education, 17*(3), 17–37.

Higgins, M., Madden, B., & Korteweg, L. (2013). Witnessing (the lack of) deconstruction: White teachers' "perfect stranger" position in urban Indigenous education. *Race Ethnicity and Education*. doi:10.1080/13613324.2012.759932

Kanu, Y. (2011). *Integrating aboriginal perspectives into the school curriculum*. Toronto, ON, Canada: University of Toronto Press.

Lather, P. (2007). *Getting lost: Feminist efforts toward a Double(d) Science*. New York, NY: State University of New York.

Lewis, B., & Aikenhead, G. (2001). Introduction: Shifting perspectives from universalism to cross-culturalism. *Science Education, 85*, 3–5.

Lewthwaite, B., & McMillan, B. (2007). Combining the views of both worlds: Perceived constraints and contributors to achieving aspirations for science education in qikiqtani. *Canadian Journal of Science, Mathematics and Technology Education, 7*(4), 355–376.

Lewthwaite, B., McMillan, B., Renaud, R., Hainnu, R., & McDonald, C. (2010). Combining the views of "both worlds": Science education in Nunavut piqusiit tamainik katisugit. *Canadian Journal of Educational Administration and Policy, 98*(1), 1–71.

Lewthwaite, B., & Renaud, R. (2009). Pilimmaksarniq: Working together for the common good in science curriculum development and delivery in Nunavut. *Canadian Journal of Science, Mathematics and Technology Education, 9*(3), 154–172.

MacIvor, M. (1995). Redefining science education for Aboriginal students. In M. Battiste & J. Barman (Eds.), *First Nations education in Canada: The circle unfolds* (pp. 73–98). Vancouver, BC, Canada: University of British Columbia Press.

Madden, B., & McGregor, H. (2013). Ex(er)cising student voice in pedagogy for decolonizing: Exploring complexities through duoethnography. *Review of Education, Pedagogy, and Cultural Studies, 35*(5), 371–391.

McGloin, C. (2009). Considering the work of Martin Nakata's "cultural interface": A reflection on theory and practice by a non-Indigenous academic. *The Australian Journal of Indigenous Education, 38*, 36–41.

McKinley, E. (2000). Cultural diversity: Masking power with innocence. *Science Education, 85*(1), 74–76.

McKinley, E. (2005). Brown bodies, white coats: Postcolonialism, Maori women and science. *Discourse: Studies in the Cultural, 26*(4), 481–496.

McKinley, E. (2007). Postcolonialism, Indigenous students, and science education. In S. K. Abell & N. G. Lederman (Eds.), *Handbook of research on science education* (pp. 199–226). Mahwah, NJ: Lawrence Erlbaum.

Menzies, C. (2004). Putting words into action: Negotiating collaborative research in Gitxaala. *Canadian Journal of Native Education, 28*(1 & 2), 15–32.

Nakata, M. (2007a). *Disciplining the savages: Savaging the disciplines. Exploring inscriptions of Islanders in Western systems of thought*. Canberra, Australia: Aboriginal Studies Press.

Nakata, M. (2007b). The cultural interface. *The Australian Journal of Indigenous Education, 36*, 7–14.

Nunavut Social Development Council. (1998). *Towards an Inuit Qaujimajatuqangit (IQ) Policy for Nunavut: A discussion paper.* Iqaluit, NU, Canada: Author.

Peat, D. (2002). *Blackfoot Physics: A new journey into the native American Universe.* Newbury Port, MA: Weiser Books.

Pillow, W. (2003). Confession, catharsis, or cure? *International Journal of Qualitative Studies in Education, 16*(2), 175–196.

Pomeroy, D. (1994). Science education and cultural diversity: Mapping the field. *Studies in Science Education, 24,* 49–73.

Rasmussen, D. (2002). Qallunology: A pedagogy for the oppressor. *Philosophy of Education Yearbook, 58,* 85–94.

Riecken, T., Conibear, M., Lyeahll, S., Tanaka, S., Riecken, J., & Strong-Wilson, T. (2006). Resistance through representing culture: Aboriginal student filmmakers and a participatory action research project on health and wellness. *Canadian Journal of Education, 29*(1), 265–286.

Sammel, A. (2009). Turning the focus from "other" to science education: Exploring the invisibility of whiteness. *Cultural Studies of Science Education, 4,* 649–656.

Steinbock, D., Pea, R., & Reeves, B. (2007). *Wearable Tag Clouds: Visualizations to facilitate new collaborations.* Retrieved from http://www.steinbock.org/pubs/steinbock-wearable clouds.pdf

Watt-Cloutier, S. (2004). *Bringing Inuit and Arctic perspectives to the global stage: Lessons and opportunities.* In Proceedings of the 14th Inuit Studies Conference (pp. 301–308). Calgary, AB, Canada: Arctic Institute of North America.

Webb, P. T. (2001). Reflection and reflective teaching: Ways to improve pedagogy or ways to remain racist? *Race Ethnicity and Education, 4*(3), 245–252.

Wheeldon, J., & Faubert, J. (2009). Framing experience: concept maps, mind maps and data collection in qualitative research. *International Journal of Qualitative Methods, 8*(3), 68–83.

CHAPTER TWELVE

Art-the-Garden: Wit(h) nessing Decolonial Teaching Beyond Disciplinary *Frontières*

CONTRIBUTORS: ELDER JUDITH TUSKY, ELDER CELINE
TUSKY (BARRIER LAKE), ELDER ANNIE SMITH ST-GEORGES
(KITIGAN ZIBI, BASED IN GATINEAU), KATY RANKIN-TANGUAY
(APITIPIWINNI), DOREEN STEVENS, (KITIGAN ZIBI), VÉRONIQUE
GABOURY-BONHOMME, NATHALIE BÉLISLE (GATINEAU).

A/R/TOGRAPHER: JULIE VAUDRIN-CHARETTE, PH.D
Faculty of Education, University of Ottawa

In opening, I wish to acknowledge
the traditional unceded land of the Anishinaabeg people I currently stand, love,
work on
as an artist, researcher and teacher of French-Canadian settler heritage.
I embrace and reassess tensions
Wherein positionality.
Is conjugated at, through, for, and wit(h),
Response-ability.

This chapter is a way to lay words on how *frontières*–whether they are disciplinary, linguistical, experiential, and, or between species–affect what we look at, how we pay attention to what we see, how we tend to relationships with, and why (and how) our teaching gets transformed. The Algonquin Lexicon proposes "*nànazikodàdiwag*–as an estranged couple" and "*pòneninidig*–at peace" (McGregor, 1994, p. 191) under the term "reconciled." Indeed, with every effort to reconcile, there are moments of tensions and moments of peace, moments of pauses, moments to speak. Languages are critical components if we are to undertake deep learning in relating to the land and to one another (Battiste, 2018; Ahenakew, 2019). Why, then, are the boundaries within languages left out of many curricular considerations and experiences within reconciliatory practices? Are we at peace or estranged in how languages teach us how to live with our interdependencies (Bastien and Krammer, 2004; Johnston, 2011)?

Relations and conversations on this topic led to my doctoral dissertation, wherein I examined a pedagogy of reconciliation through and with the Anishinaabemowen language (Vaudrin-Charette, 2020). Now assuming my role as an artist-researcher-teacher, within the tradition of a/r/tography (Irwin & DeCosson, 2004), where am I at, now passed the Ph.D gateway, yet at the beginning of new openings? Where are the Elders and colleagues I worked with, and why should I still differentiate them within institutional boundaries? I narrate and wit(h)ness (Bickel & Snowber, 2016) (my) emerging experiential knowledge my own roots, as a non-Indigenous settler-ally in training. As such, I acknowledge my own limitations in telling the stories and take responsibilities for any omissions or errors in how it is retold, despite several rounds of re-reads by all concerned parties. In considering where I am at, I think of the land I am on, who I am with and present with, and how it is intertwined with my professional location as a faculty developer. For the purpose of this chapter, I consider how such location was provoked circumvallating in the following enclosures:

> *on*: on unceded traditional land and Anishinaabeg territory
> *at*: a Sacred Medicine Garden
>
> *wit(h)*: Elders, Teachers, Artists and Students
>
> *yet*
>
> *for*: a colonial francophone academic institution?
>
> [and my own Ph.D. completion]
>
> *or* – how to we tend to deconstruct the fort and for who?

With all said and done, how do I conjugate reconciled knowledges in my teaching and learning? And what happens when it becomes first person plural, from "I" to "we"? Such questions matter now, as we are learning to inhabit spaces of coexistence, where we can be both "*nànazikodàdiwag*–as an estranged couple" and "*pònenindig*–at peace" (Figure 12.1).

Figure 12.1. Planting Semaa. [Garden at the college]. Gatineau, May 2019.
Source: Author

CONTEXT

On unceded Traditional Anishinaabeg Territory [_].
Ni manàdjiyànànig Màmìwininì Anishinàbeg, ogog kà nàgadawàbandadjig iyo akì eko weshkad. Ako nongom ega wìkàd kì mìgiwewàdj.Ni manàdjiyànànig kakina Anishinàbeg ondaje kaye ogog kakina eniyagizidjig enigokamigàg Kanadàng eji ondàpinangig endàwàdjin Odàwàng.Ninisidawinawànànig kenawendamòdjig kije kikenindamàwin; weshkinìgidjig kaye kejeyàdizidjig.Nigijeweninmànànig ogog kà nìgànì sòngideyedjig; weshkad, nongom; kaye àyànikàdj.

(Algonquin community of Kitigan Zibi and University of Ottawa.See translation here: https://www.uottawa.ca/indigenous/)

As a doctoral graduate in Ottawa, and as a Quebecois working at a college across the river, I sit across colonial and disciplinary boundaries in my learning. Land acknowledgment practices are met with resistance or enthusiasm in institutions, yet the question remains: who walks the talk? How are the talks associated with responsibilities? When and how does representation transform in the ability to take or not a responsible action?

Along with the Algonquin tradition, also present within several First Nations across the land, I pause and give *Semaa* (Tobacco). Collectively, we grief and honour the memory of Joyce Echaquan. Her death under abject circumstances on September 28, 2020, at the Joliette Hospital Center in Lanaudière, Quebec, near the Atikamekw community of Manawan. In October 2020, a brief entitled Joyce's Principle was issued by Attikamek communities and allies (https://prin cipedejoyce.com/en/index).

> Joyce's Principle aims to guarantee to all Indigenous people the right of equitable access, without any discrimination, to all social and health services, as well as the right to enjoy the best possible physical, mental, emotional and spiritual health. Joyce's Principle requires the recognition and respect of Indigenous people's traditional and living knowledge in all aspects of health.

Despite all recommendations and previous commitments (RCAP, 1996; TRC, 2015; Viens, 2019), Joyce Principle was not adopted by Quebec's Provincial Government. I give Tobacco, with the intention of grief, but also with the intention to shake roots, *de faire un tabac*, in the hope that our collective spirit strengthens and that curricular responsibility ensues. I give Tobacco, so that retelling the story of my experiences within the college system brings concrete curricular changes across the disciplines. From First Responders training to Core curriculum.

Will we tend to Joyce's Principle within the college system? Will it be a matter of representation—whether systemic racism is too strong of a word—or of a relational experience working to end the roots of such circumstances? Such questions expand beyond the scope of any academic chapter. But to address them now, I will let a story of experiencing unceded land as unseeded unfold.

At a Sacred Medicinal Garden [[]]

To assist in formulating the openings with such questions, I look within, and beyond the enclosures of the garden and a Francophone college located on the ancestral unceded land of the Anishinaabeg. The Garden, by its presence in a colonial Francophone environment, can be interpreted in so many ways; as a reparative gesture, as a moment of encounter with Indigenous Knowledge, as an act of trust from Anishinaabeg collaborators, as a way to defy systemic racism, or as a futile symbolic gesture in lieu of a deeper commitment by a post-secondary institution. In a true a/rt/ographic rendering, I leave this up to you as a reader, as I tell the stories in entangled unresolved ways, bearing witness to pedagogical possibilities and impossibilities "without the need for resolution" (Cole, 2006, p. 69). The Sacred Plants at the Indigenous garden are also part of an ontological bridge, as we are in a process of renewing pedagogy, practice, and research (Higgins and Madden,2019).

WIT(H) [+]

The Makakoose Medicine Garden is located on the unceded ancestral grounds, now hosts to Cégep de l'Outaouais, a college on the north side of the Kichissipi River. Incidentally, Kitchissipi river means the Garden River, in *Anishinaabe-mowen*. The name Makakoose was gifted by Anishinaabeg Elders Judith and Celine Tusky, from Barrier Lake, to honour their late Elder David Makakoose, during a celebration in May 2019 (Centre collégial de matériel didactique, 2020). People of Barrier Lake, Quebec, are also known as "people of the Garden" (Tusky & Tusky, 2019). Together, we have worked with teachers in the college, for the last five years, particularly in the context of a project to develop the medicinal plant garden which is discussed here. In gifting and sharing their experiences within the institutional context, they are co-contributing to a process of curriculum, but also of reconciliation and a collective healing journey.

Elder Annie Smith St-Georges is originally from Kitigan Zibi (Named after the Kitchissipi river, Garden River, in *Anishinaabemowen*). Her presence speaks of her urban experience as Anishinabe-kwe. As with Elders Judith and Celine, Elder Annie is committed to debunking mythologies maintained within the academic boundaries of college curriculum. The Elders' bravery in being present at various academic contexts is part of remembering, restorying, and providing spaces for intergenerational reparative gestures in such context (see Figure 12.2).

As I recall these moments, I am called to wit(h)nessing

> not just witnessing as the outside observing person, rather, you are fully being with and beside the other in remembrance and experience
>
> (Snowber & Bickel, 2014, p. 76).

a deeper engagement within and beyond the boundaries of my own linguistic perceptions, experiences, and limitations. In becoming with a/r/tography (Irwin & DeCosson, 2004), I reflect here on how our relationships unfolded and refolded through different times within that same place. I play as an artist, teacher, and researcher, with my own relational anchors and shifts, in experiencing geographical, historical, biological, or pedagogical boundaries. I inquire about their limitations through our resistances and their expansions through spiritual teachings contained within the plants.

One of these limitations, in the course of this chapter, has been to mediate real-life relationships with the Indigenous traditional and contemporary holders of their sacred knowledge, but also, with each contributor. The challenge is the retelling of rich experiences in ways that are not representational, but experiential; decentering narrative voice as an a/r/tographer, yet, tells that same story.

Figure 12.2. *Makizinan.*
Source: Author

Bien entendu, we would each have something to say about our experience with each other and the plants. What their disciplinary expertise brought to the Garden. Doreen is the Artist. Katy, the Social Worker. Véronique, the Biologist. Nathalie, the Geographer. Me, the Researcher. Students involved were part of Police and First Responders Programs at the college.

The garden has been a common ground to multiple openings

yet

my voice pauses with the contributors (Younging, 2018). Sweetgrass lingers as I try to let go of some disciplinary or linguistical boundaries within our relationships.

Barrière,
Jardin,

Rivière.

Les frontières de qui ?

For a Colonial Francophone Environment? [L]

Julia Ostertag (2018), settler-ally of German descent, (2018) reminds us that

> the etymology of the word 'garden' is 'enclosure.' This suggests that, for better or for worse, gardening can entangle humans and nonhumans in complex webs of words, materials, practices and beings that connect across time and space (p. 5)

As I consider my conversations with Elders as research (Kovach, 2015), the excerpts conveyed here are, in their own way, their own enclosures. As such, my intentions in writing is to consider reparative gestures (Sameshimina, Maarhuis, & Wiebe, 2019) based on self-location and responsible actions. Yet, all mistakes are mine and should be brought to my attention should they be sources of unpedagogical discomfort.

In other words, depicting the garden without identifying the complex entanglement of relationships it has contributed to seed would only contribute to an academic status quo of reconciliation through appropriation of Indigenous knowledge(s) to reinforce settler's privilege. In this sense, although I attempt to weave here a counterpoint story with a narrative of settler's guilt, heroism, or pride, the tensions between these and ethical-relational narratives are always thin threads from which we teach. In other words, in naming the fact that my retelling may be dismissed based on my positionality, I only wish to point out how such approach inhibits possibilities of dialogue.

Madden calls for a narrative linking "traditional teachings, survival mechanisms, and political engagement" (2019, p. 299). In sharing the work with Elders and colleagues, I am still sitting alone in typing this chapter. Form informs practice, and field work takes so many dimensions (Haig-Brown, 2010). Yet, this paper is only one outlet, my own narrative of the relationships folding and unfolding at the college, at the garden, and much beyond.

OR [?] HOW AND WHO DO WE TEND TO DECONSTRUCTING THE « FORT », AND WHO FOR?

As efforts towards reconciliation are becoming more present in academia, the way in which they are done, and with which relations between communities and institutions, moves forefront (Gaudry & Lorenz, 2018). Bopp, Brown, and Robb

(2018) remind us that underestimating the complexity of paradigm wars and recognizing the superficial willingness to consider power and privilege relationships remain obstacles to the processes of decolonization and reconciliation. In testifying to micro, small-scale, daily changes as well as the larger-scale changes from an institutional perspective (Louis, Poitras-Pratt, Hanson, & Ottmann, 2017).

METHODOLOGY

Revisiting the garden and its co-creation process offers a moment to contemplate entanglement in relationships with Elders, Teachers, Students, and Plants, and my emerging voice as an artist, researcher, and teacher. To make these frontiers visible, yet, to dance and transform them, I am drawn to Emberley's (2014) Indigenous testimonial uncanny as it let us disclose

> the centrality of spectatorship to imperial forms of spatial division that sever human relations with other humans, more-than-human lives, the land, and what remains unknown, or, as yet, unseen (p. 12).

In the narrative below, I began to learn to be a wit(h)ness, to listen into events with grief, relationality, joy, and peace. First, I provide methodological explanations on how I share moments were I, then we, became wit(h)ness to each other. The connections with artistic practices, plants, and research are revealed. Then, I explain the educational meanings of shifting from a representational to experiential way of teaching and learning, and how it may affect teaching in the STEAM. In my conclusions, I synthesize how such findings affect linguistical, epistemological, and pedagogical ways.

An A/r/tographic Assemblage

In a certain way, the art-making (see Figures 12.3, 12.4, and 12.5) on, at, wit(h), yet for and or the garden, has been a practice of alternative writing, as a way to "refine text by contexts outside text" (Haig-Brown, 2010, p. 927). Artistic practice has been an occasion to privilege to ponder on intentionally experiencing moments of relationality and reciprocity through wit(h)nessing as a creative and contemplative act (Walsh & Bickel, 2020, p. 140). As I mentioned above, the presence of the plants assists with attentiveness to the emergence of unexpected teaching and learning within our dialogues. Uncanny remains within the perpetual making, such as Indigenous languages unstatically alive despite colonial statistics.

The "A" in STEAM pedagogy. A pedagogy of joy and regenerative hope is central to discussions in garden pedagogy (Williams & Brown,, 2010). In fact, we may envision encounters in a garden as an extension of disciplinary boundaries. Going beyond these disciplinary boundaries constructed by the conventional norms of neoliberal driven culture of academia focusing on "progress", or exploring their nature, might recentre our pedagogies towards social and ecological sustainability

Furthermore, arts integrated educational research is based on an inductive approach that uses the elements, processes, and strategies of artistic creation as a revealing agent in a research process (Sinner, Leggo, Irwin, Gouzouasis, & Grauer, 2006, p. 1238). In my own renderings, I, too, found joy in creating an embroidery connecting the threads of conversations with all contributors, in June 2019. I reveal here an embroidered summary of our Strawberry Moon conversations, which are then translated otherwise.

As teachers and students are trying to make sense of an exponential amount of information, Kelly (2011), reminds us that "the arts act like this mighty medium that also allows for the intimate conversation between the soul of the world and the human soul: eye-to-eye, being to being, essence to essence" (p. 97). For Gardner (2018), "artistic and contemplative practices are the site of innovation and regeneration of research, teaching, and social change" (p. 58). Contemplative and artistic practices require both an attentive presence and an openness to the unknown (Walsh, Bickel, & Leggo, 2015). The threads are there. They

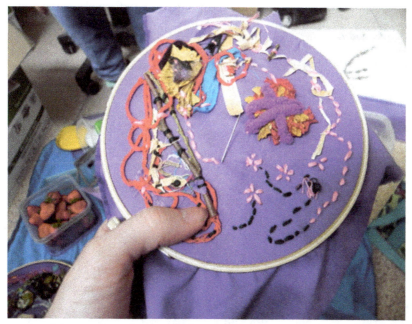

Figure 12.3. Odemiyin-Ghizis [Embroidery] June 2019.
Source: Author

bring me, as an artist, researcher, and teacher, to a form of peace preventing me from inscribing immediate meaning to words. It has enabled me to tend to relational shifts in envisioning my own epistemological, linguistic, disciplinary, and pedagogical roots.

And a bit of "M". With a thought for the scientific reader who has made it so far, I summarized those anchors as a reflexive mantra, a mnemonic to be reminded of humility, a non-equation:

$$[\backslash\_ + L = /?]$$
On, At, wit(h) languages / Or for?
Conjugating Land Language Love
And Healing

Despite being a typical artist-non-maths person, I acknowledge the beauty of formula in synthesizing complexity. In this sense, what is related here remains only one element of the complex relational process–of assemblage–of experiencing the seeding of pedagogies of reconciliation in a Francophone collegiate colonial environment. Like languages, Anishinaabemowen included (Noodin, 2015), this non-equation is a remembrance of linguistic, onto-epistemological, ethical, and pedagogical reconciliations as dialogues. Never fixed by the exercise of writing what they become in our practices. Simply, conversations and encounters wit(h) nessing moments of learning and unlearning. Some of which woven here as a story into our journey within de/colonizing.

Learning as Wit(h)ness to Grief [On Unceded Land]

On this cold winter morning, the Garden is yielding to rest. Collectively, we are grieving presence, missing the proximity of others in our teaching and in our everyday life. We are estranged to each other in the need to respect physical distancing.

I miss the contact with Elders Celine, Judith, and Annie, I miss gathering with colleague at the Makakoose Garden. Stepping by on a cold unconfined day. I am reminded of Mouse's tiny treads in the snow and her resilience in the midst of winter, when the garden is asleep.

So small, yet with her own teachings in surviving Nordic winters.

I am reminded to take small steps on my own path to de/colonial teaching.

Pondering on the year that just passed, I think of Indigenous fight for the survival of Moose in LaVerendrye Park (https://www.cbc.ca/news/indigenous/moose-moratorium-checkpoints-end-1.5769385). Of the Elders on the barricades, risking their own health and protection. I feel restless, upset about the ongoing violence against land, species, humans. Shannon, Maisy, missing Joyce,

Figure 12.4. Tiny threads in the snow. March 2020.
Source: Author

such are the insults of systemic racism. Red dress dismantled from barricade, at the other end of the country. How many voices silenced, at the edge of how many forests? For Sandra Macpherson (2018), "Sharing grief opens the present up to the past by recognizing how the affective economies of fear and hate continue to circulate" (p. 145). In these circumstances, can *Semaa* be a teacher in grieving process for reconciliation? Embroidering a tobacco pouch, I reflect on practice of responsibility, as we give before taking, may assist as we hope to develop the "patterns of practice for equity" (Cochran-Smith, Ell, Grudnoff, Haigh, Hill, & Ludlow, 2016). What if the curricular adjustments were made now as a way to transform current and past systemic fear and hate?

AN "UNCANNY" SILENCE AND A GIFT FROM ELDERS

Conjugated with the present and requires, beyond awareness, remedial action in higher education. As such, "being a witness, and being with, connects us to a history that may or may not affect us personally, but can no longer be denied" (Emberley, 2014, p. 7).

Figure 12.5. Grief and Resilience, *Kitigan Zibi*. [Embroidered Journal]. June 2019.
Source: Author

In the Fall of 2019, Elders Judith and Celine Tusky shared with me and one of the teachers (T) on their experience of separation from their families as they were picked up by government officers. An uncanny moment of grief in this dialogue comes back to Elder Celine's questioning, which invites the teacher to imagine this separation, in the present, for her, within her family. In Elder Celine's question, I hear echoes of the high rates of care and placement of Aboriginal children in non-Aboriginal families through social services, a situation denounced by the Truth and Reconciliation Committee (2015) as a continuation of colonial assimilation (Brittain & Blackstock, 2015). As I reconsidered my transcripts, of our conversations, I noted I did not ask questions, only heard with my heart wide open, and my body engaged in embroidery. Being attuned to our silences, made us present to that moment. I, as a researcher, and the teacher who was with me, became more present to what it took, for Elders to be there with us at the college, as they went back to the day they were taken from their family to attend St. Marc's residential school, near Amos, Quebec.

Elder Celine: Today the children each have their own room, their own belonging, you stop that from one day to the next, that's what happened to me.

Elder Judith: You go away to a completely different world from what you left there with your parents, the next day you don't have any more.

Elder Celine: We were small enough that we didn't even know which direction we were going. I was lost there.

Teacher: You didn't know where you were.

Elder Celine: I wasn't able to. All I remember is that the bus was pretty ugly, an old bus.

Teacher: It was coming to pick you up.

Julie: The school bus.

Elder Celine: We had nothing, we were almost naked. I don't even remember putting on my shoe, I was looking for my running shoe, I was looking for my vest, I was cold.

Elder Judith: We hadn't been notified that they were coming for us, not even a warning, no one knew.

Teacher: They were coming early as well, they were coming to get you. My God, your mother must have been shocked. Do you remember that?

Elder Celine: Like it was yesterday.

Elder Judith: As we said in the circle in Maniwaki, I've never forgotten what I went through ...

Teacher: On my God.

Elder Judith: Some of them stayed there for years, but they're far away now.

        [silence]

Teacher: I can't imagine ...

Elder Celine: If it had existed now, what would you be doing today? If the law were to return to the way it was before, in every family?

Teacher: If they came to take my children and then left? I can't even imagine ...

Elder Celine: Yes, how would you react?

Teacher: Anger, I don't see any other words, helplessness, anger.

Elder Judith: even with anger, you can't do anything.

Teacher: You're becoming emotionally, completely ... lost.

Julie: Unbearable [long silence]

For Emberley (2014)

> To create reparative practices through testimonial and storytelling epistemologies means agreeing that historical wrongs and traumas do not simply remain in the past but give rise to how people experience present-day realities (p. 17)

Upon their return to Rapid Lake, after experiencing the trauma of residential school, Judith and Celine learned traditional medicines and language through their Elder, David Makakoose. As such, in bringing *Semaa* at the Cegep, they proposed that the garden should be named in honour of Elder David Makakoose, honouring the transmission of ancestral traditional knowledge, from the community of Rapid Lake, through their knowledge, words and presence.

In this sense, we may look at the garden as the very beginning of a "reparative gesture where the creative builds a relation" (Sameshima et al., 2019, p. 156). Does the resilience of this knowledge, transmitted to the Elders upon their return from the residential school, materialize in the experiential relationship with the plant world? Planting *Semaa* at the college, I become a wit(h)ness, and with it, ponder on plants as ontological bridges. A student stops to ask me: "Isn't it strange to give tobacco to a garden that grows tobacco?" I am grateful she knows about *Semaa*, and that the gifts of Tobacco's works from within.

AT THE SACRED GARDEN: SHARING DISCIPLINES

The increase in allied postures among teachers and in the pedagogical accompaniment of reconciliation at the college level cannot be done without the active presence of Elders and Indigenous Community members within academia. Yet, such presence can lead to increased pressure on communities, be delegitimized, or tokenized by a growing number of institutions. However, remembering that campus learning gardens are often at the margins in the context of the neoliberal university (Ostertag, Gerofsky, & Scott, 2016), helps us put in perspective our individual and collective experiences of transformation.

TEACHING STEAM WITH SEMAA

As with several reconciliation initiatives within academic contexts, the presence of Anishinaabeg Elders and partners at the college has been tied to various invitations from Francophone teachers committed to developing a relationship of respect and reciprocity in collaborations (Scully, 2018; Pidgeon, 2014). In relation to STEAM, *Semaa* implies reconsidering past and current knowledge within and beyond their disciplinary allegiances. For example, future police officers become artists, biologists coordinate ecosystems of care, Elders and community calls on collegial accountability. The Garden assists in bringing gratitude to the forefront when reconsidering our relationships as Elders-teacher-student-advisor, past, current, future.

In fact, it has been pointed out that transdisciplinary discussions rarely include a revision of the epistemic and pedagogical assumptions implied within each discipline. Using a hermeneutic approach, Yeo (2016) unpacks the "disciplinary elements related to openness to questions, the interpretation of experience, the nature of knowledge, our relationship to language and text, and being-in-the-world" (p. 50). If we find some reference points of disciplinary allegiances in the following excerpts from our conversations with the teachers, we also see conciliations of their relationship to knowledge, that expand beyond the enclosure of their disciplines.

> Nathalie: The way of living in nature is experiential all the time and this is what brings you to your wisdom. The ways of teaching of the Elders influenced my perspective … to trust your life experiences and the knowledge that you have accumulated over time,

which is part of you. To trust that is the case for the students. To trust that TEK knowledge is also evolving through time, in constant motion. (Belisle, 2019)

Veronique: For me, the garden was really about sharing the harmony of nature, [and] nature is primordial to our survival. For me, that's the first thing, then the fact that we respect the territory, and the fact that the Elders welcome us, that we have a relationship, that's what's important to me. [...]

(Gaboury-Bonhomme, 2019)

As seen in the above comments, within the context of the college, the garden emerged from a relational space for meeting, acknowledgment of the territory and of Anishinabeg language and knowledge. These elements correspond to certain elements identified by Bopp et al. (2018) as factors contributing to a pedagogy of in-depth reconciliation, namely "working in communities of practice with Aboriginal stakeholders and learning in action" (p. 7). Yet, teachers, Elders, students, and myself, are still confronted with several *colonial boundaries* (Donald, 2012) in our experiences at the college. Can the medicines and us humans, be part of countering the "monocultures of thoughts" (Cole, 2006, without being instrumentalized? Hence, it is with this Tobacco—and understanding the protocol, that opened teachers' perspectives on teaching and learning beyond their disciplinary worldview. Which in turn lead to focusing on diverse relationships/connection that exist across space and time (human and non-human; past and present).

WIINGASH: JOYFULLY BRAIDING TRANSDISCIPLINARITY AND RECIPROCITY

In this sense, looking at transdisciplinarity in the garden allows us to locate relations within the STEAM equation. As we should see in this unfolding story, the cross-pollination of the arts, the garden, and teaching sciences and technical traits at the college, the contribution of the aesthetic experiences also plays a role in encountering other ways to relate within, and transcend beyond, the colonial boundaries of the disciplines.

Wiingash (Sweet grass) and her sweet gift of reciprocal relations (Kimmerer, 2013) takes me to the lived curriculum and the relational world of teaching. Sweet grass holds a promise for embracing reciprocity as a responsibility (Kuokkanen, 2011). In a way, caring for a garden requires, too, a presence. Such presence implies, at times, an attentiveness to fears, trauma, experiences that taught us how schools failed entire generations of Indigenous children and communities. Beyond the enclosure, research on garden as pedagogy illuminates the connections with mindfulness, love, attention, and joy (Dilafruz, 2015). For Hauk (2019) "reframing garden processes as embodied and living nexus [...] open up

Figure 12.6. Imprints. [Garden's mosaic by Animitagize Odaying – Clear water woman speaking from the heart (D. Stevens)]. Gatineau, May 2019.

Source: N. Belisle, 2019, with permission

new possibilities" (p. 8). In this sense, *Wiingash* provokes our experiences in the garden as expressions of an embodied experience of interconnection and relations. Beyond the enclosure, perhaps we can see how we may be accountable for the stories of this time.

TEACHING AND LEARNING: SHARING JOY AS A WAY TO RELATE

Aesthetic Joy. To look at this issue in a synthetic way, I bring to story an experience of co-creation of joy at the garden. In creation, *poiesis* is a moment of passage, in which artist and participants co-create a moment of ecstatic transformation (Sameshimina et al., 2019). An Anishinaabeg artist is present that day to install her mosaic work. Two students, possible future police officers, had volunteered to provide logistical support for our event. They are invited to join the artist in fixing mosaics to the stones, to create art in the garden (Figure 12.6).

In our subsequent conversation, Doreen talks about this experience an "epiphany of joy":

> When the girls came to help, first they were working at the practical/medical things, measuring and pouring, and being patient and helping each other as a good team. Finally, when they figured "oh, we are really enjoying ourselves", they started working on this rock or this other one, and I said go and do it yourself, because it is your hand, it is your magic, your own magic.

I just showed you the ingredients, then you print it yourself on the rock with your own hand, your own magic and your own time and this is a memory. So you are printing a memory there that cannot be changed, something that was like an epiphany of joy, because they were so relaxed and they said "I really like this".

(Stevens, June 2019)

She continues the importance of teaching with love, *Sàgìhidiwin* in Anishinaabemowin. The heart is the first teacher. In projects with Indigenous youth, transferable skills of creation transposed to other life situations, such as planning and story sharing for life learning (Victor et al., 2016). Holistic learning connections are also reflected in Katy Rankin-Tanguay's experiences as Indigenous Cultural Agent at the college, conversing about her training practices:

When giving a cultural workshop, I explain that we have an oral culture and a different way of teaching and learning. I ask them to try something different: To step out of their mind, to listen with their hearts, not just their ears, like when we listen to our Elders. Try not to take notes if possible. We believe that our mind, our spirit will take only what's needed for today. It is a more natural way of learning for us. At that point, I see some relief, some smiles among the students. (, personal communication, 2019)

The practice of co-creating an Aboriginal garden at the college, initially outside the curriculum, became a safe space, to explore what Haig-Brown (2010) referred as "deep learning experiences within Indigenous knowledges" (p. 286). As one of the teachers participating in the workshop summed up, the ecological and gestural encounters with the Anishinaabeg artist collaborator provided further educational impressions, as "the tactile imbued their learning with something concrete and it summed up what they had learned" (Gaboury-Bonhomme, 2019).

WIINGASH AND STEAM TEACHING

Louis et al. (2017) reminds us that "it is through gifting and sustained time in making that fabricates a softened and loving relation with the other, changing the possibilities for new communal relationships" (p. 156). In reflecting of the student's experience of joy in the act of creating, I reflect on the experiences of security and or vulnerabilities within colonial academic institutions. Does the practice of creation, both individual and collective, manage to transpose this feeling? On the contrary, does it help us become more aware of our vulnerabilities, our privileges, and responsibilities?

All these conversations, their iterations, and impressions, bring me back to Vicky Kelly (2014), to her invitation to cherish and respect the essence of what moves our hearts.

> We embrace things of the world, which we take into ourselves holding them as unique
> and precious. We take them to heart, impressing them there, each in its own unique way
> abiding in our hearts; there each one echoes, resounds, and resonates […] This imagina-
> tive metaphor revealed a living truth to me and I sensed that although I could perceive it,
> it lived in essence free of my ability to perceive it (p. 93)

Wiingash has opened up our perspectives to transdisciplinary, in witnessing our
own resistances and allegiances to art, natural, or social sciences. In installing the
mosaics at the garden, the students were present to a moment of encounter, and
trust, where they, too, became creators. The artist and contributors' comments on
teaching reveals their personal engagement in questioning the power/privilege
relations present in the educational milieu (Davis et al., 2018). As a researcher,
bringing my a/r/tographic practice (Springgay et al., 2008) into academic walls
was also an experience in reconsidering reciprocity (Vaudrin-Charette, 2020).
What might this mean for welcoming Anishinaabe Elders and the presence of
the Anishinaabemowen language within a college environment?

KIISHIK: HUMILITY AND ACCOUNTABILITY

With *Kiishik*, the gift of humility meets engagement, academic disciplines, and
questions accountability in our intentions in teaching. I steep with Cedar's teachings and infuse with the ideas of protection, and care, through time.

> A cedar will grow up as it grows down. One can see this for oneself on those shorelines
> where erosion has washed away the soil. One can see the shape of the cedar under the
> earth, how it reproduces in kind the exact shape of the cedar above the surface. If a cedar
> is twenty feet tall, it will also have twenty-foot roots. The roots will be in the same posi-
> tion as the branches, a mirror image of them, in fact (Geniusz, Makoons, 2015, p. 37)

Interestingly, in the technical (Eurocentric) world, cedar-leaf is explored as a
"high-performance symmetric supercapacitor" (Wang et al., 2018, p. 2046).
Returning to the knowledge of Elders in their traditional teachings on plants
corresponds here to an ecological conception of reconciliation "challenging the
challenges of not reproducing Eurocentric epistemologies that separate us from
our relational links with the environment" (Butler et al., 2018, p. 30). For Elder
Judith Tusky the teachings of cedar (kiishik) and humility are the first steps
towards reconciliation. Kiishik, is here a witness and included in the deployment
of the meetings of elders, collaborators, and plants in the colonial collegial context, and of our entanglement with the natural world (Zembylas, 2018).

CEDAR AND HUMILITY IN STEAM

The dimension of the relational link to nature puts disciplinary knowledge into perspective and brings us back to the process of reconciliation. In the words of Anishinaabeg Elder Annie Smith St-Georges,

> People are not conversing anymore, they are sleeping, and they are not watching nature. This is a playground for squirrels, they are not here, where are they? People only see nature as a resource and not as part of being alive. People cut the tree down and they don't know that it has a heartbeat. On tree roots, there are microscopic mushrooms that may trigger the growth of a certain flower two hundred feet further away. This is why we call on all our relations. People separate everything. They go into their house and do their own thing. There is no communities. If you want to decolonize, if everyone is stuck in their house and their computer, how will people understand what is harmony with nature and with each other? (Smith St-Georges, 2019)

As Elder Annie shared, we can appreciate our current challenge, on a global basis, to return to the humility of being in nature in the midst of social isolation increasing our felt dependency on technology.

Another "uncanny moment" occurs during the conversation with Elder Annie, imbued with the gentleness of *kiishik* in her invitations to be humble.

> It is hard. You are taking on a deep project, but it is needed. It is a process that has been ongoing for a while. It has been 150 years since J.A. McDonald, now we have to go back and undo it. So when institutions come to me, I say it has to be an open process, the archives have to be accessible, the Elders have to be recognized.

> [*A cedar branch flies over and land on my notes*]

> It may be that we have to go back to the Two Rows Wampum. I don't know how you're going to share this in your PhD.

> Julie: I will share it with you. Miigwech, Kokom Annie.

In conversation with Elders and Kiishik, responsibility remains both collective and individual. Yet, to teach responsibility is to let one learn how he, she, they can be able or recognize their inability to respond. Thus, I am now left with a question—how am I currently accountable for my ongoing commitment to support Indigenous languages revitalization, and blooming? Am I honouring the People of the Garden and their Anishinaabemowen language in reciprocal ways? To reflect on this question, I turn to my conversations with Elder Celine and Judith,

Elder Celine: The contribution for the garden, I tried to be available, sometimes it was not easy, because of the distance and the weather. I learned myself for me to meet with the teachers and see what they want to know. Everyone was

	very respectful for us to come here to share, very welcoming here. And … I don't know if you remember some of the words we taught you?
JVC:	… *Debewin* [Truth] … *Zaagdhiwin* [love] … and there is a really long one … respect.
Elder Celine:	*Manadjiiwewin*, that is respect.
JVC:	the ones that I remember more is the plants, *Semaa, Wiingash, Mshkode-washk, Kishiik*.
Elder Judith:	*kishkaatik*, kishiik is *Maniwaki* we have different dialogue from back home, than Kitigan Zibi (KZ).
Elder Celine:	so when I use this dictionary, it is from KZ. It is not all the same. So I should use the writing from your community, the one that you write?
Elder Celine:	yes, this one is the way they write from the priest. When you go to Church, when they give you a book to sing. They all write the same way as the priest.
Elder Judith:	every anishinaabemowen, it is the same background, so we understand each other.
JVC:	yes, the aki. For the Garden, we will have the panels with the Four Medicines and the land acknowledgement, and I do not know, would we want to have audio as well? Because if it is written differently, from one place to another, is it better?
Elder Judith:	For me I think it better to frame it, to see it with the plants, it would be good, in French, English, Anishinaabemowen.
JVC:	Now we do not have the translation in Anishinaabemowen that is why we did not put it up now, because we want to see all three languages at once.

This short excerpt presented above bears witness to my own colonial biases.

Unpacking Frontiers Reassessing Our Linguistical Allegiances

In the above conversation, I was dealing with writing conventions derived from relationships with religion; with the presentation of the Anishinaabemowen language in the college in audio or written form, which remains embryonic; and with the challenges encountered in memorizing the language, despite a commitment to it. I relinquish the authority postures of Eurocentric research to be present, and with it, my vulnerabilities. Madden (2019) reminds us that reconciliation cannot be separated from "the persistence of colonial power relations and the injustices they continue to produce" (p. 300). Reconciliation requires a continuous reflexion on our own biases, a first and ongoing conciliation of what lies with our intentions and efforts to reconcile.

Fels (2014) invites us to reflect that "our arrival and possible renewal requires that natality be an embodied presence of reciprocal responsibility and mindful awareness and attention to who we choose to be in the presence of each other" (p. 120). Here, I continue grapple with a question—How, in these circumstances, can we become allies of the possibilities of Aboriginal languages at the college level? How can we reconcile our epistemologies beyond our practices, in all our relationships?

DISCIPLINARY MSHKODEWASHK: SPIRITUAL BOUNDARIES AND STEAM

Is it taboo to touch on the spiritual world, or is it yet another boundary of the fort? What does this mean for Indigenous knowledges in academia, and, in particular, for STEAM education?

Mshkodewashk brings the wisdom of sacred spaces and psychology and the ethical transformations held within the co-creation of the garden. Sage and its teachings emerge during discussions with participating teachers. Here, the natural science teacher talks about reconciliation of the spiritual dimensions of teaching by sharing her experience of smudging with Sage during a meeting with the Elders and students.

> I could be told that right now we are in a secular state and you have no business giving them a smudge. But it's not [a] religion, it is spirituality, but for some people it can be seen as a ritual that is associated with a belief. It's complex,
> At the same time, I use it because it's soothing, it's scientifically proven that sage has chemical components that calm us.

> (Gaboury-Bonhomme, 2019)

These statement from the natural science teacher shows the tension between the validity of knowledge and its integration into experience. In her own work of her own work of understanding the relationship of diverse knowledges, the teacher remains confronted with disciplinary boundaries by exploring the links between practices of Annishinabeg and the scientific method. She emphasizes that the consideration of the spiritual aspect remains ambiguous in her teaching, despite its integration into personal experience. Her narrative allows us to situate some of the vulnerabilities and relations of authority in the negotiation of knowledge relationships in a pedagogical context of reconciliation.

There had been multiple exploration done by "outside" scientists on the properties on Sage. *Salvia Apiana*, is another given name for Sage following scientific nomenclature. Its properties have been unveiled in biochemistry (Afonso et al., 2019), along with potential use as an antimicrobial agent (Cordova-Guerrero et al., 2016) and diabetic treatment (McCuan, 2013). Moreover, the case about the fundamental distinctions between First Nations' traditions of smudging as spiritual beliefs and religious practices in schools was also made in a historical case involving the provincial government of British Columbia and Nuu-chah-nulth Tribal Council (Hunter, 2020).

The outside eye may try to interpret, but it is provoked by the engagement of other senses, including, as we all experienced in pandemic times, our senses of smell and touch. This is quite clearly observed in current deprivation of tactile sensory experiences within online learning environments. Yet, as people learn smudging practices in their own environment, they may welcome it widely in schools, college, and universities, with the help of Smudging protocols,

> Smudging allows people to stop, slow down, and become mindful and centred. This allows people to remember, connect, and be grounded in the event, task, or purpose at hand. Smudging also allows people to let go of negative feelings and thoughts. Letting go of things that inhibit a person from being balanced and focused comes from the feeling of being calm and safe while smudging. (Government of Manitoba, 2019, p. 4)

> *kihêw waciston*, which means "eagle's nest" in Cree, is a home away from home for Mac-Ewan University's Indigenous students. Our ceremonial and gathering room is available to students who want to smudge on campus. *kihêw waciston* staff can provide you with sweet grass or sage, or you are welcome to use your own medicines from home. Whether you want to smudge on your own or with someone else, we welcome you to our space. (MacEwan University, kihêw waciston, n.d.)

Sage brought together so many diverse voices. I wit(h)nessed the teacher's process in reconciling the scientific and spiritual aspects of sage by bridging its benefits through both biochemical and spiritual experiences. I also heard various responses to how smudging is perceived, welcomed, or experienced in schools and academic institutions. I am yet again left with a question: Is burning sage within institutions expanding and creating more spaces for Indigenous students and Indigenous knowledges, thereby enhancing the well-being of the whole community (beyond the fort)? Can we collectively co-create caring and supportive environments for all learners, in our dialogues between cultures, languages, disciplines, and species?

SYNTHESIS: ART-IN-THE GARDEN AS ONTOLOGICAL BECOMINGS

In my note-making, plants and wools move around on the canvas, always becoming, and still transformed as they become part of the embroideries. Yet, as they are, effectively, no longer where they belong, how can I, effectively, acknowledge both their "presence and absence" (Aoki, 2005)? The gesture is not only mine, and I am becoming part of this ontological bridge, slowly threading. The circle

	Estranged---- Reconciliation---At peace I -------------We	
Linguistic	Interpretation Representational	Sustained Experiential
Epistemological	Distant	Interdependent
Disciplinary	Coded disciplines	Expansion
Pedagogical	Teacher centred	Student centred
Tends to?	Trauma	Resiliency

Table 12.7. Pondering onwards ...

Source: Author

becomes a frame where I see and relate differently to the teachings of plants and land. Here, I am reminded by the Elder Judith's teaching, "It is not just a word, Anishinaabemowen, for us it is the whole picture" (Tusky, 2019).

In relating the stories and their counterpoints offered by the Elders' conversations, I attempt to thread together "traditional teachings, coping mechanisms and political engagement" (Madden, 2019, p. 299). These 'uncanny moments' along with the wit(h)ness of the plants offered space of sharing filled with grief, joy, silence, embodied experience. Ultimately, these uncanny moments led to the co-creation of knowledge and exploring ways to reconcile the division of knowledge, which in turn led to the implementation of the garden. Now in time, the garden stands, and is itself a witness of life, past, present, future. I wanted to bear witness to the ethical (collective contributions) and pedagogical conciliations of co-responsibility of the learning community. At the heart of the creative process and the silences noted, epistemological meeting points emerged. Traditional and current knowledge included in the Anishinaabemowen language, the stories of resilience of the Elders present at the college, and an embodied experience of our tensions and vulnerabilities allowed us to co-create, with *Semaa*, *Kishiik*, *Wiingash*, and *Mshkodewashk*, a path to reconciliation.

EPILOGUE

In my conclusions, I synthesize how such findings affect linguistical, epistemological, disciplinary, and pedagogical ways (see Table 12.7).

The celebration of the Garden brought together community members, teachers, students, and staff members. We all gathered to gift *Semaa* to offer our gratitude, and infuse paths of reconciliation into the garden, surrounded by *Kiishik*, for protection. "The rain will help Semaa to seep into the ground, in rooting our intentions", Kokom Celine said.

I gaze at the forest, as the college sits at the edge of a natural preservation area, with paths and trails where humans regenerate, breathe in, and meditate. Ten years ago, the same forest was a theatre for violence, when a young woman went missing, her body found, later, by others. Her voice silenced. With so many Indigenous women and girls missing in this country, how can red dresses on barricades keep being dismantled? How do we live, inhabiting simultaneously the unspeakable and the healing paths on this land?

The Algonquin Lexicon proposes "*nànazikodàdiwag*–as an estranged couple" and "*pònenindigi*–at peace". How I choose to walk with these teachings on, at, with, for, and or relies on,

embodied healing
heartful resilience

silently resting

respectfully landing

With the Acknowledged teachings of Traditional Sacred Plants:
Tobacco, Cedar, Sweetgrass and Sage. In memoriam:
Joyce Echaquan 1988–2020.

ACKNOWLEDGMENTS

Miigwech to: Elders Judith and Celine Tusky, Elder Annie Smith St-George, Doreen Stevens, Katy Rankin-Tanguay, Nathalie Bélisle, and Véronique Gaboury-Bonhomme and to all students how took part in the becoming of the Makakoose Garden.

REFERENCES

Afonso, A. F., Pereira, O. R., Fernandes, Â. S., Calhelha, R. C., Silva, A., Ferreira, I. C., & Cardoso, S. M. (2019). The health-benefits and phytochemical profile of Salvia apiana and Salvia farinacea var. *Victoria Blue Decoctions. Antioxidants, 8*(8), 241.

Ahenakew, C. (2019). *Towards scarring our collective soul wound.* Musagetes: Creative Commons. Retrieved from https://decolonialfutures.net/towardsscarring/

Aoki, T. T. (2005). Toward Curriculum in a new key (1978/1980). In W.F. Pinar & R.L. Irwin (Eds.), *Studies in curriculum theory: The collected works of Ted T. Aoki* (pp. 89–110). Mahwah, NJ, US: Lawrence Erlbaum Associate Publishers.

Bastien, B., & Kremer, J. W. (2004). Blackfoot ways of knowing: The worldview of the Siksikaits-itapi. Calgary, AB : University of Calgary Press.

Belisle, N. Settler. (2019). Lives in Gatineau, QC. Personal communication. June 2019.

Bopp, M., Brown, L., & Robb, J. (2018). *Reconciliation within the Academy: Why is indigenization so difficult? Teaching commons.* Lakehead University, Sudbury, ON. Retrieved from https://teac hingcommons.lakeheadu.ca/sites/default/files/inline-files/bopp%20brown%20robb_Recon-ciliation_within_the_Academy_Final.pdf

Brittain, M. et Blackstock, C. (2015). First Nations Child Poverty : A Literature Review and Analysis, First Nation's Children Action Research and Education Services (FNCARES). Retrieved from: https://fncaringsociety.com/sites/default/files/First%20Nations%20Ch ild%20Poverty%20-%20A%20Literature%20Review%20and%20Analysis%202015-3.pdf

Butler, J. K., Ng-A-Fook, N., Forte, R., McFadden, F., & Reis, G. (2018). Understanding Eco-Justice education as a praxis of environmental reconciliation: Teacher education, indigenous knowledges, and relationality. In G. Dans Reis & J. Scott (Eds.), *International perspectives on the theory and practice of environmental education: A reader* (pp. 19–31). Cham: Switzerland, Springer.

Centre collégial de matériel didactique, CCMD. (2020). *Décolonisation de nos relations, de nos pratiques et de nos environnements éducatifs.* Retrieved from https://www.ccdmd.qc.ca/nouvel les/2020/decolonisation-de-nos-relations-de-nos-pratiques-et-de-nos-environnements-ed-ucatifs

Cochran-Smith, M., Ell, F., Grudnoff, L., Haigh, M., Hill, M., & Ludlow, L. (2016). Initial teacher education: What does it take to put equity at the center? *Teaching and Teacher Educa-tion, 57,* 67–78.

Cole, P. (2006). *Coyote and Raven Go Canoeing: Coming home to the village* (Vol. 42). Montreal, QC: McGill-Queen's Press-MQUP.

Cole, P., & O'Riley, P. (2010). Coyote and Raven talk about equivalency of other/ed knowledges in research. In P. Thomson et M. Walker (Eds.), The Routledge doctoral student's compan-ion: getting to grips with research in education and the social sciences (p. 323–334).Oxford-shire, UK: Routledge.

Cordova-Guerrero, I., Aragon-Martinez, O. H., Diaz-Rubio, L., Franco-Cabrera, S., Serafin-Higuera, N. A., Pozos-Guillen, A., … & Isiordia-Espinoza, M. (2016). Antibacterial and antifungal activity of Salvia apiana against clinically important microorganisms. *Revista Argentina de microbiologia, 48*(3), 217–221.

Davis, L., Hare, J., Hiller, C., Morcom, L., et Taylor, L. (2018). Challenges, possibilities and responsibilities: Sharing stories and critical questions for changing classrooms and academic institutions. *Canadian Journal of Native Education, 40*(1). Vancouver, BC: University of British Columbia.

Dilafruz, R. W. (2015). *Regenerative hope: Pedagogy of action and agency in the learning Journal of Sustainability Education,* Vol. 10, November ISSN: 2151–7452.

Donald, D. (2012). Indigenous Métissage: A decolonizing research sensibility. *International Journal of Qualitative Studies in Education, 25*(5), 533–555.

Emberley, J. V. (2014). *The testimonial uncanny: Indigenous storytelling, knowledge, and reparative practices.* New York, NY: SUNY Press.

Gaboury-Bonhomme, V. Settler. Lives in Gatineau, QC. Personal communication. June 2019.

Gardner, Morgan. (2018). Relations To Live By. *Artizein: Arts and Teaching Journal, 3* (1), art.10. https://opensiuc.lib.siu.edu/atj/vol3/iss1/10

Gaudry, A., & Lorenz, D. (2018). Indigenization as inclusion, reconciliation, and decolonization: navigating the different visions for indigenizing the Canadian Academy. *AlterNative: An International Journal of Indigenous Peoples, 14*(3), 218–227.

Geniusz, M. S. (2015). *Plants have so much to give us, All we have to do is ask: Anishinaabe Botanical Teachings.* Minneapolis, MN: University of Minnesota Press. Project MUSE.muse.jhu.edu/book/43058.

Government of Manitoba. (2019). *Smudging protocol and guidelines for school divisions. Minister of Education and Training.* Winnipeg, Manitoba: Indigenous Inclusion Directorate. Retrieved from https://www.edu.gov.mb.ca/iid/publications/pdf/smudging_guidelines.pdf

Fels, L. (2014). Woman Overboard: pedagogical moments of performative inquiry. In B. Bickel, C. Leggo, & S Walsh (Eds.), *Arts-based and contemplative practices in research and teaching* (pp. 112–123). New York: Routledge. Retrieved from https://doi.org/10.4324/9781315813387

Haig-Brown, C. (2010). Indigenous thought, appropriation, and non-Aboriginal people. *Canadian Journal of Education, 33*(4), 925–950.

Hauk, M. (2019). When the garden becomes the campus: Five strategies for just sustainabilities, Prescott College and the Institute for Earth Regenerative Studies, Paper for the Interactive Symposium, *University Learning Gardens: Cultivating the Margins, on Borrowed Land and Time.*103rd Annual Meeting, American Educational Research Association. Toronto, Canada.

Higgins, M., & Madden, B. (2019). Refiguring presences in Kichwa-Lamista territories: Natural-cultural (re) storying with Indigenous place. In C.A. Taylor & A. Bayley (Eds.) *Posthumanism and higher education* (pp. 293–312). Rotterdam, The Netherlands: Springer.

Hunter, J. (2020, January 8). Smudging ceremony in school did not violate freedom of religion: B.C. Supreme Court. *Globe and Mail.* Victoria, B.C. and Nuu-chah-nulth Tribal Council. Retrieved from https://www.theglobeandmail.com/canada/british-columbia/article-smudging-ceremony-in-school-did-not-violate-freedom-of-religion-bc/

Johnston, B. (2011). Th!nk Indian: Languages are beyond price. Neyaashiinigmiing, ON: Kegedonce Press.

Kelly, V. (2014). To See, To Know, To Shape, To Show: The path of an indigenous artist. In S. Walsh, B. Bickel, & C. Leggo (Eds.), *Arts-based and contemplative approaches to research and teaching: Honoring presence* (pp. 45–65). New York, NY: Routledge Books.

Kelly, V. (2011). Finding Face, Finding Heart, and Finding Foundation: Life Writing and the Transformation of Educational Practice. Transnational Curriculum Inquiry, 7(2), 82–100.

Kimmerer, R. W. (2013). *Braiding Sweetgrass: Indigenous wisdom, scientific knowledge and the teachings of plants.* Minneapolis, MN: Milkweed Editions.

Kihewwaciston, MacEwan University. (n.d.). Retrieved from https://www.macewan.ca/wcm/CampusLife/kihewwaciston/CommunityandCulture/index.htm

Kovach, M. (2015). Indigenous methodologies: Characteristics, conversations, and contexts. Toronto, ON: University of Toronto Press.

Kuokkanen, R. (2011). *Reshaping the university: Responsibility, Indigenous epistemes, and the logic of the gift.* Vancouver, BC: UBC Press.

Louie, D. W., Poitras-Pratt, Y., Hanson, A. J., & Ottmann, J. (2017). Applying indigenizing principles of decolonizing methodologies in University Classrooms. *Canadian Journal of Higher Education/Revue canadienne d'enseignement supérieur, 47*(3), 16–33.

MacPherson, S. (2018). *From spectator to citizen: Urban walking in Canadian Literature, performance art and culture.* Unpublished doctoral thesis, Faculty of Arts, Department of English. Ottawa, ON: University of Ottawa.

Madden, B. (2019). A de/colonizing theory of truth and reconciliation education. *Curriculum Inquiry, 49*(3), 284–312. Retrieved from https://doi.org/10.1080/03626784.2019.1624478

McCune, L. M. (2013). Traditional medicinal plants of indigenous peoples of Canada and their antioxidant activity in relation to treatment of diabetes. In *Bioactive food as dietary interventions for diabetes* (pp. 221–234).

McGregor, E. (1994). *Algonquin Lexicon.* Kitigan Zibi, QC: Kitigan Zibi Education Sector.

Noodin, M. (2015). Weweni: Poems in Anishinaabemowin and English. Detroit, MI: Wayne State University Press.

Ostertag, J. A. (2018, fall). *Pathways. The Ontario Journal of Outdoor Education, 31*(1).

Ostertag, J., Gerofsky, S., & Scott, S. (2016). Learning to teach environmental education by gardening the margins of the academy. In D. Karrow, M. DiGiuseppe, P. Elliott, Y. Gwekwerere, & H. Inwood (Eds.), *Canadian perspectives on initial teacher environmental education praxis* (pp. 186–216). Ottawa, ON: Canadian Association for Teacher Education (CATE).

Pidgeon, M. (2014). Moving beyond good intentions: Indigenizing higher education in British Columbia universities through institutional responsibility and accountability. *Journal of American Indian Education, 53(*2*)* 7–28.

Royal Commission on Aboriginal Peoples. (1996). Report of the Royal Commission on Aboriginal Peoples (RCAP). Retrieved from https://www.bac-lac.gc.ca/eng/discover/aboriginal-heritage/royal-commission-aboriginal-peoples/Pages/item.aspx?IdNumber=400

Sameshima, P., Maarhuis, P., & Wiebe, S. (2019). *Parallaxic praxis: Multimodal interdisciplinary pedagogical research design.* Wilmington: Vernon Press.

Scully, A. (2018). Whiteness and land in indigenous education in Canadian teacher education. [Unpublished doctoral dissertation]. Lakehead University, Thunder Bay, ON.

Smith St-Georges, A. (2019). Kitigan Zibi First Nation. Lives in Gatineau. QC. Oral teaching. Personal Communication, June 2019, Gatineau, QC.

Snowber, C., & Bickel, B. (2015). Companions with mystery: Art, spirit, and the ecstatic. In S. Walsh, B. Bickel, & C. Leggo (Eds.), Arts-based and contemplative practices in research and teaching: Honoring presence (pp. 67–87). Oxford, UK: Routledge

Sinner, A., Leggo, C., Irwin, R., Gouzouasis, P., Gouzouasis, P., & Grauer, K. (2006). Arts-based educational research dissertations: Reviewing the practice of new scholars. *Canadian Journal of Education/Revue Canadienne De l'éducation, 29*(4), 1223–1270.

Springgay, S., Irwin, R. L., Leggo, C, & Gouzouasis, P. (2008). Being with A/r/tography. Rotterdam: Sense Publishers.

Stevens, D. (2019). Kitigan Zibi First Nation. Lives in Ottawa, ON. Oral teaching. June 2019, Ottawa, ON.

Tusky, C., & Judith, J. (2019). Algonquin Nation of Barrier Lake. Rapid Lake, Quebec. Oral teaching. Personal communication. May 2019, Gatineau. QC.

Tusky, C., & Judith, J. (2019). Algonquin nation of Barrier Lake. Rapid Lake, Quebec. Oral teaching. Personal communication. November 2019, Gatineau. QC.

Truth, & Reconciliation Commission of Canada. (2015). *Canada's Residential Schools: The final report of the Truth and Reconciliation Commission of Canada* (Vol. 1). Montreal, QC: McGill-Queen's Press-MQUP.

Vaudrin-Charette, J. (2020). Une étude a/r/tographique de la présence de l'Anishinaabemowen dans un collège francophone au Québec. [doctoral dissertation]. Retrieved from https://ruor.uottawa.ca/handle/10393/40815

Victor, J. Linds, W., Episkenew, J., Goulet,L., Benjoe, J., Brass, D., Pandey, M., Schmidt. (2016). Kiskenimisowin (self-knowledge): Co-researching wellbeing with Canadian First Nations youth through participatory visual methods. *International Journal of Indigenous Health, 11*(1), 268–275. DOI: 10.18357/ijih111201616020

Viens, J., (2019). Commission d'enquête sur les relations entre les Autochtones et certains services publics: écoute, réconciliation et progrès. Rapport final. Sous la présidence de l'Honorable Viens, J. (dir.). https://www.cerp.gouv.qc.ca/fileadmin/Fichiers_clients/Rapport/Rapport_final.pdf

Walsh, S., & Bickel, B. (2020). The Gift of Wit(h)nessing Transitional Moments Through a Contemplative Arts Co-Inquiry. Canadian Journal for the Study of Adult Education, 32(2), 137–154.

Walsh, S., Bickel, B. & amp; Leggo, C. (2015). Arts-based and contemplative practices in research and teaching: Honoring presence. Oxford, UK: Routledge

Wang, Y., Yang, D., Lian, J., Pan, J., Wei, T., & Sun, Y. (2018). Cedar leaf-like $CuCo2O4$ directly grow on nickel foam by a hydrothermal/annealing process as an electrode for a high-performance symmetric supercapacitor. *Journal of Alloys and Compounds, 735*, 2046–2052.

Williams, D.R. & amp; Brown, J.D. (2010). Living soil and composting: Life's lessons in the Learning Gardens. Clearing Magazine, 2010 Compendium Issue: 40–42.

Yeo, M. (2016). Decoding the disciplines as a hermeneutic practice. *New Directions for Teaching and Learning, 150*, 49–62.

Younging, G. (2018). *Elements of indigenous style: A guide for writing by and about indigenous peoples (Indigenous collection)*. Edmonton, AB: Brush Education.

Zembylas, M. (2018). The entanglement of decolonial and Posthuman perspectives: Tensions and implications for curriculum and pedagogy in higher education. *Parallax, 24*(3), 254–267. Retrieved from https://doi.org/10.1080/13534645.2018.1496577

An Axiology for Making– Weaving Slow Pedagogies with Indigenous Pedagogies–First Peoples' Principles

LORRIE MILLER & SHANNON LEDDY
University of British Columbia

One braid of wool from merino sheep sits atop a table in my home studio, smooth, soft and white as a cloud; the other, also merino, dyed into silvery grey, red, along with a hint of charcoal. The blended results remind me of an exotic granite. I resisted breaking up the fibre jewel for months, satisfied with just looking at it and feeling its softness. Inevitably it would either be felted or spun into a knit or woven project.

(Miller, 2019, personal reflection)

INTRODUCTION

In this chapter we describe two versions of a co-created workshop that blends together the tenets of *slow* and Indigenous pedagogies, exploring the pedagogical implications for this form of collaborative teaching, learning, and investigation in an effort to support the infusion of Indigenous learning across curricular areas. At times in our narratives, one voice will take the forefront, while the other steps back. We share this writing along with discussions here, neither one taking "lead" authorship, but sharing in our scholarship.

We two are visual art educators and teacher educators who are collaborators, colleagues, and friends. We bring our greater identities to our work as a Métis scholar (Shannon), and a settler scholar who is birth mother to two Cree sons and also two children with settler roots (Lorrie). Both of us hailing originally from Saskatchewan, and having made our homes in Vancouver for the past several decades. We bring our "ways" of being in the world—as educators, researchers, artists, mothers, women, humans, and embrace *slow* and Indigenous pedagogies to inform our practices (FNESC, 2008; Payne, 2005; Pinar, 2010; Volk, 2015; Lane et al., 2012). The workshop we created has been delivered at an international conference to arts educators, researchers, and students, to a graduate class at Simon Fraser University, at a Western Canadian teacher educator and teacher candidate conference, and to a salon of scholars and educators who focus on early childhood learning. These workshops, each in their own context, were designed to illustrate the connections to Indigenous pedagogies in this collaborative and integrated approach to *making*. This chapter recounts the events from two of these workshops. We also reflect on our experiences to illuminate the value of slowing down and taking learning as an organic and lived pedagogy, fulsome, and with room for the emergence of what might be possible.

We are colleagues in a faculty department focussed on curriculum and pedagogy who had known one another for some time before we found ourselves working at the same place. Shannon teaches Indigenous education courses through the lens of contemporary arts and Lorrie teaches arts education through her work in textiles classes. Through our conversations on the philosophy and mechanics of teaching and learning, we began to notice significant overlaps in our respective work with Indigenous and slow pedagogies. Our work together began one afternoon in June, 2019, when we met to create two scarves, both woven and felted from roving, as a timed trial-run in preparation for an upcoming workshop at an art educators conference. With our workshop limited to just one hour, we had much to do. Felting, we'd decided, would be the best way to illustrate our ideas. Wool, an amazing material, is both sustainable and ethical when produced locally through careful tending to the wooly animals, whether they are sheep, goats, or camelids. Many questions began to be unpacked with students, teachers, artisans, and artists. Why not a synthetic *wool*? What is the difference between a protein fibre and an acrylic, for example? Good questions, but we need to look at the fibre structures themselves to fully answer that. The short answer is that the complexity of the fibres of wool and silk allows them to hook into each other when felted to create a new fabric out of the strands. No synthetic does this as well as protein-based fibres. Just look at them both under a microscope and the differences begin to come into focus (see Figure 13.1).

The long history of working with wool is embedded in cultures in many parts of the world. Over many centuries people have developed ways of spinning and weaving animal locks into fabrics. This action itself links us to our ancestors. For me, Lorrie, I learned to knit when I was about twelve, and have knit ever since. When looking at my family tree, my ancestors all came from cold places–Celts

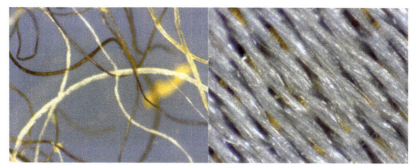

Figure 13.1. Polyester (left) & Wool (right) under magnification.
Source: Lorrie Miller

from Ireland, Scotland, Scandinavian from Sweden and Norway–wool-work was and still is prevalent. I was curious to learn more.

The patterns, techniques, and the machinery involved with the processing and production of wool and cloth are intriguing. The invention and proliferation of the Jacquard loom and its subsequent impact on computing illustrates a clear link to engineering, math, and design, along with production in the Western context. The automated Jacquard loom, with its use of punch-cards to determine complex patterns in woven cloth allowed for more weavings, greater distribution, and faster production. Such rapid development in technology was not brought into practice without controversy or resistance. The proliferation of the mechanical loom brought about significant change to local artisans and weavers that relied on their cottage industries to support their families. Other similar weaving innovations have been present in non-Western societies for far longer. Flax weaving, for example, has been dated back to at least 5000 BCE in Egypt, and there are examples of basket work in Peru dating back to 8000 BCE, and textile work dating to 700 BCE (Stone-Miller, 2002). Modern textiles created by Pacific Northwest Coast First Nations weavers remain highly prized for their complexity of design. Whether the mathematics inherent in the process of weaving is culturally articulated or not, mathematical functions and processes are certainly present in the process of creating fabric and patterned fabric.

The terms we hear a lot nowadays in the field of education, including cross curricular learning, interdisciplinary projects, collaborations, and cooperative efforts—all of these ideas and actions existed long before STEM, and now STEAM, were dreamed into existence. Our current and growing interest in officially making connections across the disciplines within sciences and arts is not without controversy and raises many questions. Can one really bring art into science without ample knowledge of the discipline? Is one discipline simply used as the handmaiden of the other? Will the disciplines become watered down or diluted? Is creativity the exclusive province of art only? Perhaps the root of these questions lies in the Western ontological notion of separability, the idea that real understanding occurs by breaking things down to their constituent parts. This approach, rooted in scientific thinking that is necessarily anthropocentric, is sharply at odds with Indigenous ways of knowing that are holistic and focused on interconnectedness in which everything (animate and inanimate alike) is equal and plays a part. "While perspectives are changing, there are notions of humans as being separate from the natural world within certain disciplines of Western science" (Michell, 2018, p. 47). Here, we challenge these persistent silos and the Western ontology that underpins them.

Through our work we aim to deconstruct the colonial gaze, that silent arbiter of self and other, rooted in the oil paintings of exotic Others and vast tracts of "terra nullius", and in curiosity cabinets from previous centuries (Bell, 1982). This gaze framed Indigenous peoples in the Americas as both exotic and less-than for harbouring understandings of land that were not rooted in ownership or development. It fuelled the fires of Manifest Destiny and expansionist policies that were aimed at offering land, a scarce and expensive commodity in the European context, to newcomers who could then define themselves as equal to the landed gentry they left behind. We wish to disrupt the "monarch-of-all-I-survey[1]" illusion of ownership (Pratt, 2007, p. 205) that simply maintains a colonial status-quo, to inform shifts in pedagogical practices for art educators, cross-curricular teachers, and researchers. Part of our hope is to illustrate that a lived and experiential learning, as messy as it might be, reflects both inner and outer learning—in our case, through producing a felted and woven wool scarf that is an *artifact* of the *learning*, not the goal of the workshop.

We selected wool weaving and felting as the embodiment of our metaphor of merging practice and pedagogy for several reasons. Have you ever tried to felt fleece into felted fabric by hand—quickly? It cannot be done. Throughout the wet and soapy process, one may sweat, roll the bundle of wool, and unroll, check—roll and continue, but this takes time and *it takes the time that it takes* for the fibers to latch onto one another and become a singular textile. This process of co-creating a singular work was intended to be more than a *how-to* make a scarf in a felting workshop. It was a *why-to make*, and what does it mean to *make* together in a workshop.

In considering how to illuminate Indigenous pedagogical approaches we chose to work with the First People's Principles of Learning (FPPoL) (FNESC, 2008; see Figure 13.2), which we know is in wide use in classrooms throughout our province. Collaboratively developed by the First Nations Education Steering Committee, they are not attached to the teachings of any particular nation, but rather point to some collective understandings about relationships, interconnectedness, and holistic sensibilities that inform the ontologies of many nations around the province and across the country. Although not strictly a set of pedagogies, this document offers some important pedagogical considerations for working with Indigenous students and for including Indigenous voices and content. The experiential mode of the workshop and co-construction of knowledge also model Indigenous pedagogies as we work together.

In order to structure our thinking and workshop discussions, we were drawn to the holistic cast of the Medicine Wheel as a framework for our metaphorical weaving. This is another pedagogical tool that is not connected to any particular nation, but which is taken up by several broad cultural groups including the Cree, Anishinaabe, and Métis nations. Although many of the coastal and interior nations in British Columbia where we live use different models to describe holistic perspectives on selfhood, we appreciate this model drawn from the prairies for its connections to First Nations and Métis peoples in other parts of Canada, and to the needs of mixed urban Indigenous populations.

Each workshop had two distinctive aspects. In our Medicine Wheel work, based on the writings of Lane et al. (2012), participants were asked to join us in situating their learning within the four quadrants of the Medicine Wheel (spiritual, emotional, physical, and intellectual), to explore the ways this might inform and influence pedagogies. This aspect also includes a discussion of the FPPoL and how each principle relates to the holistic framework of the Medicine Wheel. The hands-on aspect of the workshop involved a group making process where participants worked around a single long table. Together we co-constructed a scarf—a dual metaphor for co-creating knowledge as well as a visual artifact of that knowledge. From our teaching experience we know that by offering participants a physical piece of our collective weaving, we are sending them away with a mnemonic reminder of the learning we collectively built together.

Our workshop illustrated that learning through the interdisciplinary lens of STEAM is not strictly about discipline integration when understood as a holistic approach through an Indigenous lens, over time, with materials that are local, inviting reflection and storied learning. This is a lived pedagogy. Our approach to teaching in this context was not top down, but rather rooted in student experience, drawing on the knowledge of participants with prior textile knowledge. Our discussion of the First People's Principles of Learning asked participants to plumb their own learning experiences for examples that illustrate each principle from their own perspectives. We aimed to help them tap into the pedagogical affordances of the FPPoL in their own approaches to teaching and learning. In this way, we collaboratively explored and constructed understandings together. We also brought in the tenets of "slow", not in a tracking of time and completion of a task, but a deeper and broader concept, possibly best known through the popularization of the slow-food movement (Petrini, 2007; Slow Food International, 2020).

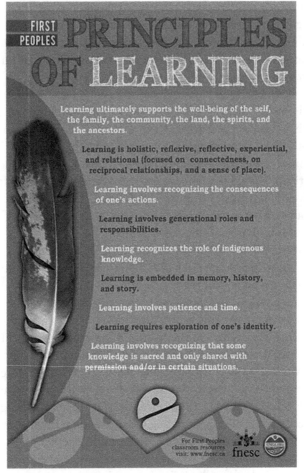

Figure 13.2. First people's principles of learning Poster. FNESC (2008)
Source: Reproduced from FNESC (2008)

"Slow" Impacts on Co-constructing Knowledge and Pedagogy

Since the inception of this movement, sparked in 1986 by Carlo Petrini, other related philosophies and practices have emerged, including, but not limited to: slow science, reading, travel, journalism, fashion school, and specifically slow pedagogy. All of these thoughts and practices have inspired critical works in the field of Education including *The Slow Professor* (Berg & Seber, 2016). Here we drew from common overarching tenets of "slow", and specifically the slow cloth principals that Lipson (2012) identified as elements within process, culture, materials, and soul. Lipson (2012) notes that within a slow process, appropriate time is used for a task to develop skills to the level of mastery. Rushing is clearly not a part of this idea, and volume is not the goal, but rather the quality and connections. Cloth and culture cannot be separated, and multiplicities of cultures inform our practices. Slowing down is an important part of connecting ourselves and our students to our collective and respective ancestors and intergenerational practices, to our descendants, across to our peers, and within other practices where we may influence one another. We are part of many communities, and slow-cloth values can foster a community where we can learn and teach one another.

This slowness connects deeply with Indigenous pedagogies and practices where learning often happens intergenerationally, and through stages, where each new element of what is being learned builds upon what is already mastered and understood, and what is connected to the practices of living within a community. Tewa scholar Gregory Cajete (1994) described the process of slowing down and making connections as learning nested in concentric circles. This element of Indigenous pedagogy reminds us that when we take learning slowly, we are better able to see the holographic web of interconnections to all aspects of life, and all types of life. Lipson (2012), likewise, with *her Slow Cloth Manifesto*, links the practices of making with textiles to contemplation, creative expression, joy and beauty, along with careful consideration of materials, the sustainable use of resources, appropriate selection of materials, and the pleasure brought about by making, using, and sharing of textiles made mindfully.

From this perspective of teaching and learning, we emphasize the need to know the materials and concepts of one's inquiry, and one's own creation. Through our own practices of teaching and learning, and within these workshops, we have learned:

(a) about the materials we use, and here in particular about wool, the nature of wool, how it acts and behaves, how it can be manipulated, and about the local wool industry and ethical considerations for sourcing materials;

(b) about our own identities as the weavers, felters, and co-creators of an object. We ponder together how we are changed by this collaborative and collective engagement;

(c) that a collaborative knowledge approach feeds into realms of our other professional work- artistic practices, our teaching, our scholarship;

(d) that symbols and metaphors help us to understand more than the written word that we privilege for knowledge expression. This approach also requires us to relinquish the notion that learning can ever be complete; that all of the learning in say, for example, grade three curriculum occurs only within that year; that a terminal degree signals the end of something. Learning is reiterative and recursive.

Below we describe the specifics of two workshops where the visual, textural, and lived experience of learning are attended to through the dual lens of slow and Indigenous pedagogies. We hoped to bring more than a "how-to" approach to our participants and dip into notions of "why-to."

WORKSHOP 1: INSEA CONFERENCE, UBC, JULY 2019

In July 2019, our home university, The Faculty of Education at the University of British Columbia, hosted the World Congress for the International Society for Education through Art. The theme of the conference was "Making." Scholars and art educators from all over the globe gathered to explore notions of making today in pedagogy, research, and artistic practices. All workshops were scheduled in an art lab designed for making of all sorts (see Figure 13.3).

We set up in our assigned alcove within the textiles quarters, sharing space with a wool spinner. A donated stack of burlap sacks containing freshly shorn wool added the distinctive aroma of barn and earth to our corner. The wool was *unskirted* and full of earthen matter and vegetation; it was nowhere ready to be felted and was relegated to materials for processing for the fall term, but not now. We set out our supplies: bins for water, sponges, olive-oil soap, two blue pool noodles cut into two-foot lengths, old beach towels, lengths of bubble wrap, and a matching length of door-screen, the kind readily available at any hardware store. Though these are all common materials, for our participants this was a new way to use each of them. The one unfamiliar material, the central object of "making" is the wool roving, a merino hybrid with silk with blended hues of green and blue and grey. The material itself is distinctly beautiful, shimmery, rich in colour, soft to touch and delicate. Loose fibres cling to damp fingers, leading to a playful exploration. The day was sunny and we could see the central green path from the full height windows glazing the full northern wall of this art lab. We were the

Figure 13.3. INSEA Weaving and Felting Workshop at UBC.

Source: Authors

second session of the day and in competition with many other alluring activities and talks.

Our workshop included an introduction to the medicine wheel as a pedagogical framework, and a guided exploration of these ideas so participants could begin to see how concepts from "Western" pedagogical approaches mesh with Indigenous pedagogies. The goal was to dip into the ideas, and to allow time to settle into the concepts long enough to grasp a depth and breadth ripe for further exploration. This was not an activity easily rushed.

During the workshop a group of strangers gathered around a single table with materials already laid out. The task was a simple one, on the surface; collaborate to weave wool roving together into a plain-weave (over-under) scarf, and then use a wet-felt technique to cajole the fibres together. We opened the session with an acknowledgment that our work was being done on the traditional and unceded territory of the Musqueam people, and though a familiar introduction, we add to this by offering each participant the opportunity to state not only their name, but their relationship to the land they had come from. These histories complicate our relationship with the land. Our stories, narratives, positionalities, link us together, here and now, as we embrace learning that is lived, shared, and remembered.

Within the confines of an hour, we led this group through two linked activities, one of making, and one of reflection, analysis and sharing. The weaving itself, both the action of weaving and the resultant object, for us is a metaphor for slow and Indigenous pedagogies with the warp and weft of the same cloth is stronger together, and then when felted, its fibres lock in with each other becoming inseparable in a single fabric.

Once the scarf was finished, solid and mostly dry, we invited participants to cut a swatch of the scarf to take with them as an artifact of their learning, and to thread a small wooden pony bead onto a fringe while placing an intention into that artifact and bead. We asked them to consider this question: *What will you take with you from this experience into your teaching and artistic practices?* After the whirlwind hour of making and reflecting, we tidied up and reflected on next steps for us. More time than one hour, we determined was needed. As well, rather than dividing the group for activities with the Medicine Wheel work (see Figure 13.4), or the felting, and then switching back, we decided that next time we would stay as a single group engaging in each activity together, trying to adhere to a time constriction without appearing to feel rushed. Feeling rushed seems to lead to acting rushed, which in turn may well lead to rushing others. We hold to the idea that learning takes as long as it takes.

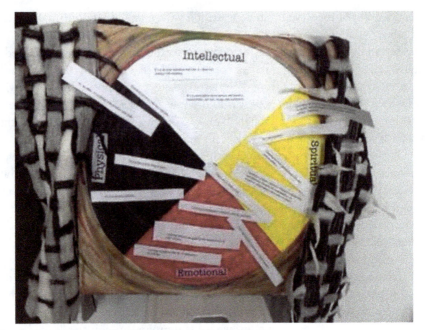

Figure 13.4. Medicine Wheel Framework and Felted Weaving.
Source: Authors

WORKSHOP 2: SIMON FRASER UNIVERSITY GRADUATE COURSE, NOVEMBER, 2019

Perhaps one of the greatest pleasures of teaching in a post-secondary institution, particularly in a Faculty of Education, is working with graduate students. Often these students are in-service teachers who have returned to school with the genuine desire to develop and improve their teaching practice, and to expand their perspectives with a view to new career options. And because, as hooks (1994) tells us, "to teach in a manner that respects and cares for the souls of our students is essential if we are to provide the necessary conditions where learning can most deeply and intimately begin" (p. 13), it becomes both a delight and a challenge to offer such students learning opportunities that are at once rich for them, and at the same time, transferrable to their students.

The workshop that we developed presented just such an opportunity to a group of MEd students working in a community-based cohort with a focus on ecological education at Simon Fraser University. During our final session together, I (Shannon) invited Lorrie to come and share our work with students. In the context of this particular course, which focused on Indigenous pedagogy, we built on many of students' prior learning experiences, including other Indigenous guest speakers, meeting with Elders at the UBC Indigenous garden, attending a powwow at the Squamish Nation in North Vancouver, and viewing works by a number of Indigenous artists.

We arrived in class early on a Saturday morning with nearly all of our materials at the ready. After beginning with an opening circle that gave students a chance to check-in and learn a little about Lorrie, we provided an overview of the process and some of the key ideas we would cover. Although I had previously introduced students to some basic elements of the medicine wheel, we reviewed the four components of holism that form the back of the teachings associated with this framework: intellectual, spiritual, emotional, and physical. Our goal was not to be prescriptive about unpacking or applying the framework, but rather to invite students into a relationship with it that would allow them to connect each step of their learning that morning, and indeed from the whole class, with each of the quadrants.

The fifteen students in the class self-selected into two groups whom we then set the task of preparing two long tables at which each group could work (see Figure 13.5). I knew the students were tired. We'd met five times over the course of the term, for four or five hours on a Friday evening, and then for another eight hours the following day. It was a rather intense schedule, given that all but one of the students was a full-time teacher as well. But there was palpable excitement and a lot of chatter as we laid out lengths of fine mesh screen and bubble wrap along each table and fetched the bins of warm water we would need to get our work done.

After Lorrie gave an initial orientation to the process we were about to unfold together, each group was invited to explore the many different colours of roving we'd brought along and to choose those they'd like to work with. Once each group made their selections, we each attached ourselves to a group and began to gently coach them as they worked the roving into the different lengths required (long for the warp, short for the weft) and began to develop a pattern for their weaving. Lorrie's group chose three colours, a beautiful variegated red along with a more natural grey and a creamy white, while my group chose four, a bright green, a bright red, a rich aqua, and the same creamy white as the first group (see Figure 13.6). We invited the groups to share with one another how and why they'd made their selections.

While many hands make light work, too many bodies around a table can prohibit it, so at this point we invited students to partake in the second aspect of the workshop, which was to work with the First Peoples Principles of Learning (FNESC, 2008; see Figure 13.2) and the medicine wheel. We have been using the medicine wheel (Lane et al., 2012), as an organizing framework for pedagogy that attends to the whole learner, emotional, spiritual, physical and mental,

Figure 13.5. Collaboration in process.
Source: Shannon Leddy

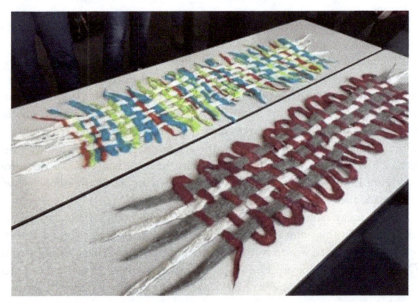

Figure 13.6. Results of felting workshop 2.
Source: Lorrie Miller

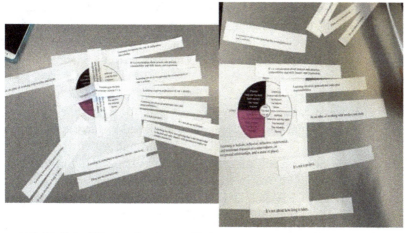

Figure 13.7. Medicine Wheel and pedagogical linkages.
Source: Shannon Leddy

while relating this to FPPoL and slow pedagogies (see Figure 13.7). Prior, within this workshop, we had the whole group work together with four printed copies of the wheel, and four copies of the FPPoL, along with adages drawn from the *Slow Stitch Manifesto* (Lipson, 2012). Each principle was cut out into a strip of paper and students were asked to work together to determine which principle they felt worked with which quadrant. Some students gravitated quickly to this activity, while others clearly preferred to keep their hands busy with the many stages of weaving and felting. In the end, of course, everything got done and it was clear that students enjoyed the experience tremendously and took great pride in their work.

At the start of our work, with so much going on in the room, it was a little hard to corral all that energy, and I felt a bit anxious about time, knowing that we had more work to get through that afternoon as well. Despite knowing that things take as much time as they take, especially in a technical process such as felting, my busy little brain is in the habit of constantly casting forward, and I must often call my own attention back to the task at hand. In many ways, I always appreciate the irony that this work presents a significant challenge to my natural tendency to want to get through everything quickly. Perhaps that is why I love this type of teaching so much. After a few minutes of working and chatting with students, I was quickly in the zone of flow (Csikszentmihalyi, 1997). Time anxiety slowly dissipated, as did the feeling of herding cats, and I became increasingly attuned to the work itself, and attendant to the many conversations unfolding in the room.

While most of the folks in the first such workshop we held at the InSEA conference the past summer were strangers to us, the students in this class were very familiar to me. I had observed them collectively and individually enough to be sure that this workshop was having precisely the impact we intended it to. There was new learning that was hands-on and tactile; there were some intellectual tasks, and there was an opportunity to draw on prior learning, merging it with the activities we'd organized for the session.

One of the greatest challenges we face within teacher education relates to how Teacher Candidates (TCs) perceive and plan to engage the First Peoples Principles of Learning. These principles were designed as a reflective framework for teachers to consider as they developed lessons and activities. They form a holistic pedagogical approach that articulates key values in Indigenous ways of learning and knowing, rooted in the land and in the interconnectedness of all things and of all ideas. For those new to this discourse, the FPPoL is often taken up as something teachers will share and discuss with their students. While this approach isn't wrong, per se, neither is it the best use of these concepts as a tool in teaching and learning.

In the class, the medicine wheel activity that supplemented the co-constructed weaving asked students to do two things. They were asked to consider each principle and decide on which quadrant of the medicine wheel it best fit. At the same time, they were asked which aspects of the work we did together met the criteria laid out in that principle. Further, they were asked to think of examples from their own teaching practices that met the criteria as well. In this way, students were invited into a holistic perspective on the FPPoL that revealed to them both how they already employed Indigenous pedagogies and offered them further insight into how they might do so more often and better.

CONCLUDING THOUGHTS

In our reflection on these workshops, a key idea has emerged, which is that Indigenous pedagogy is located at the intersection of content and curricular context. It is at the core of how we think about *what* we will do with students, and in *how* our teaching connects with all of the other teachings we have offered. For example, Language Arts was used throughout the entirety of the lesson because we spoke in English and provided handouts containing detailed written instructions. Content related to mathematics is evident in measuring quantities (weight, size, and volume), designing patterns, and designing workspaces. Connections can be made to social studies and science curricula through looking at characteristics of cultural examples, and in addition, considering elements of sustainability.

While it is beyond the scope of this chapter to articulate the connections across each curricular area, we invite the reader to reflect on these connections, to determine for themselves if they can see a rationale for these intersections, especially linked to curricular requirements. Further, we invite the reader to consider that if Indigenous education is understood as the centre of both content and context, what might this mean for one's personal relationship to teaching with First Peoples' Principles of Learning as a set of considerations that can inform our pedagogies for working with Indigenous students? How might this temper the inclusion of Indigenous voices and content in meaningful ways? Our hunch is that some of the anxiety that often infuses such considerations may begin to dissipate. Indigenous pedagogy considers students and their lives as being at the centre of all teaching and learning. While we recognize that this is of particular importance to Indigenous learners, we know this holds great value for *all* students. Once teachers begin to engage in decolonized ways of considering pedagogy, it becomes much more natural to form relationships with Indigenous peoples, with the land, and curricular content.

We want to be very clear though, that in offering this bridge between slow and Indigenous pedagogies, as expressed through First Peoples' Principles, we do not mean to imply that one can merely approach the work of teaching holistically and consider the work of Indigenous education done. That would be to cling to frontier logistics—an over-simplification that condemns the concept of FPPoL to the realm of words, not deeds and maintains the fiction of separate realities (Donald, 2009). Rather, we offer this twinned pedagogy as a way to come into relationship with Indigenous peoples, histories, and lands to bring new life to our curriculum and practices.

In British Columbia, the curriculum emphasizes big-ideas learning, student-centred approaches and capitalizes on the porousness of traditional curricular silos, suggesting that separability is not as useful as it once seemed (British Columbia Ministry of Education, 2019). Generally, BC's curriculum encourages a natural integration approach rather than a forced marriage of unlike disciplines, honouring the principle that nothing exists in isolation. For example, we use language arts and graphic representations to explicate complex scientific phenomena. Similarly, we can explain colour variation in a painting by considering how physics informs the appearance of colour. Everything is interrelated. Learning and teaching, we contend, are ultimately human endeavours, and as such are inherently emotional processes. Everything is in relationship to everything else. Embracing that notion invites the kind of paradigmatic change we work for.

NOTE

1 Pratt's notion of 'monarch-of-all-I-survey' was in reference to the 19th century landscape paintings commissioned by landed gentry to capture the vastness of their land possessions as a kind of a visual survey.

REFERENCES

British Columbia Ministry of Education. (2019). BC curriculum. British Columbia, Canada. Retrieved from https://curriculum.gov.bc.ca/

Bell, L. (1982). Artists and empire: Victorian representations of subject people. *Art History*, 5(1), 73–86.

Berg, M., & Seeber, B. K. (2016). *The slow professor: Challenging the culture of speed in the academy.* University of Toronto Press.

Cajete, G. (1994). *Look to the mountain: An ecology of indigenous education.* Durango, CO: Kivaki Press.

Csikszentmihalyi, M. (1997). *Finding flow: The psychology of engagement with everyday life.* New York: Basic Books.

Donald, D. (2009). Forts, curriculum, and Indigenous Métissage: Imagining decolonization of Aboriginal-Canadian relations in educational contexts. *First Nations Perspectives*, 2(1), 1–24

First Nations Education Steering Committee. (2008). First peoples principles of learning. Retrieved from *http://www.fnesc.ca/wp/wp-content/uploads/2015/09/PUB-LFPPOSTER-Principles-of-Learning-First-Peoples-poster-11x17.pdf*

hooks, B. (1994). *Teaching to transgress.* New York: Routledge.

Lane, P., Bopp, J., Bopp, M., Brown, L., & Elders. (2012). *The sacred tree* (4th ed.). British Columbia: Four Winds International.

Lipson, E. (2012). The slow cloth manifesto: An alternative to the politics of production: Textile society of America Symposium Proceedings. Digital Commons @ university of Nebraska – Lincoln. Retrieved from https://digitalcommons.unl.edu/cgi/viewcontent.cgi?article=1710&context=tsaconf

Michell, H. (2018). *Land-based education: Embracing the rhythms of the earth from an indigenous perspective.* Vernon, BC: J.Charlton Publishing.

Payne, P. (2005). "Ways of doing," Learning, teaching and researching. *Canadian Journal of Environmental Education*, (10) Spring, 108–124.

Petrini, C. (2007). *Slow food nation: Why our food should be good, clean and fair* (Clara Furlan & Jonathan Hunt, Trans.). New York: Rizzoli Ex Libris.

Pinar, W. (2010). Currere. In *Encyclopedia of curriculum studies*. Thousand Oaks: Sage Publishing, 178.

Pratt, M. L. (2007). *Imperial eyes: Travel writing and transculturation.* London: Routledge.

Slow Food International. (2020). *Our history – About us – Slow Food International.* [online] Retrieved from https://www.slowfood.com/about-us/our-history/

Stone-Miller, R. (2002). *Art of the Andes: from Chavín to Inca.* London: Thames and Hudson.

Volk, S. (2015). Paragraphs take time; Conversations take time. Centre for Teaching Innovation and Excellence. Oberlin College Retrieved from https://steven-volk.blog/2015/10/04/paragraphs-take-time-conversations-take-time/

CHAPTER FOURTEEN

Final Thoughts: Relational Education, Radical Hope, and Action

EUN-JI AMY KIM & KORI CZUY

Learning from the stories and wisdoms shared in this book, We have yet again been reminded of the importance of relationality in teaching and learning. The journey of this book project took many years (2–3 years) of building and sustaining relationships. At some point in time, we had to let go of some relationships.

In the forward, Dwayne mentioned the continual effects of colonial frontier logics in relationship denial as the impetus to create hierarchy of knowledge systems. Enlightenment-based knowledge systems and coming to know nature through logic and reasons are important.

We came to the term—Relational education—to describe the journey of learning from the land, teaching and learning through a relationship-centered approach. Relationships come with responsibility, without hierarchy, but with respect.

Relationships, especially those within education and science have been predominantly imbalanced and therefore based within skewed power-dynamics, making them unsustainable.

Relational education reconnects with these lost relationships, learning beyond objective, standardized, sanitized knowledge, but alongside the subjective spirit and embracing the knowledge that thrives beyond these restrictive parameters.

By suggesting some calls to action, we are not here to give any sentiments of "this is the way to go"- As humans (being educated within a colonial system), we are beginning to re-understand the depth of knowledge the Land, animals, and plants can teach us, but have often neglected our responsibility *to* them. Humility is being open to the songs, vibrations, and scientific wisdom of the Land and animals, our responsibility is to listen to them.

How can we live today so to learn from the mistakes of our past, to allow for our future ancestors to thrive, and live alongside the Land, our neighbours and community, and all within the Circle of Life through the true intentions of Treaty, through reciprocal relationships. Challenge yourselves to re-engage with, learn from, and innovate through the scientific knowledge and methodologies that thrived on these Lands for thousands of years, this is radical scientific hope! By taking seriously the wisdom of relational understandings of the world, the fine contributions provided in this book serve to create such condition.

Humility in the context of relational STEAM education for us is about first acknowledging the dual nature of educators as a lifelong learning learner at the same time a teacher. Throughout conversing with the contributors of this book, we have been reminded that teaching is a sacred profession that started from thousands of years ago. Yet, teaching in a relational approach, the teacher is always a learner. In the context of STEAM education, We urge educators and educational researchers who walked through the journey with us in this book to think about the pedagogical ways to continue the work of relational repair and renewal between diverse ideas, peoples and between human and non-human beings that is promoting balance and harmony.

Notes on Contributors

Dr. Myrle Ballard is an Assistant Professor, in the Faculty of Science, University of Manitoba. Dr. Ballard has a Ph.D. in Natural Resources and Environmental Management. She is First Nation Anishinaabe from Manitoba and a fluent speaker of Anishinaabe mowin. Her current research focuses on developing frameworks regarding Indigenous Science and Western Science, specifically on developing baseline monitors using Anishinaabe mowin for water management. She combines various tools to combine technology with oral tradition and history to capture the management systems that are embedded within Anishinaabe mowin.

Lisa Lunney Borden is a Professor of mathematics education at St. Francis Xavier University in Canada and holds the John Jerome Paul Chair for Equity in Mathematics Education. Having taught 7–12 mathematics in a Mi'kmaw community, she credits her students and the community for helping her to think differently about mathematics teaching and learning. She is committed to research and outreach that focuses on decolonizing mathematics education through culturally based practices and experiences that are rooted in Indigenous languages and knowledge systems. Lisa teaches courses in mathematics education and Indigenous education.

Rorger Bosiher was born in New Zealand but set sail for Canada in 1974. In New Zealand, he comes from Ngati Kahungunu (Hawke's Bay) but his current project there is a biography of Selwyn Murupaenga—Ngati Kuri painter, actor, orator and playwright. In Canada, he has Musqueam friends and neighbours and is a UBC Professor Emeritus in Adult Education. For many years, he has researched learning in out-of-school settings in China and been at the forefront of UNESCO efforts to theorize and build learning villages, districts, towns and cities. Boshier helped create the boisterous (and naughty) Shuang Yu learning village in China— next to a river where Confucian traditions meet modernity.

Julie Vaudrin-Charette recently completed her PhD at the Faculty of Education, University of Ottawa. She works as a Curriculum and Faculty Developer at Cegep de l'Outaouais, and is a strong advocate and practitioner of decolonizing and equity educational practices. Julie's a/r/tographic research dwells into how deep learning and Indigenous languages are experienced in pedagogies of reconciliation. She is a mother of four beautiful humans and accompanying them in their becoming.

Dr. Kori Czuy, Γ"dΛ"∇ʔ°, is Métis/Polish with Cree ancestry, and was born in Treaty 8 by the banks of the Peace River. She is the Indigenous Engagement Specialist at the TELUS Spark Science Centre, focusing on supporting community to bring forward multiple ways of knowing science. Kori is on an ongoing journey to reconnect with and learn from the knowings of the land, and helping others connect with the complexities of these knowings alongside Western science. She has a PhD in storying mathematics; through her research she worked with children and Treaty 7 Elders to explore the depth of mathematics within local Indigenous stories.

Dwayne Donald, is a descendent of the amiskwaciwiyiniwak (Beaver Hills people) and works as a professor in the Faculty of Education at the University of Alberta. He is Canada Research Chair in Reimagining Teacher Education with Indigenous Wisdom Traditions, and his work focuses on ways in which Indigenous wisdom traditions can expand and enhance understandings of curriculum and pedagogy.

Amanda Fritzlan is a doctoral candidate in the Department of Curriculum and Pedagogy at the University of British Columbia. Her research engages with practicing teachers' experiences of connecting with place and community through mathematics education. She is particularly interested in methods of narrative inquiry. Amanda has taught extensively in the K-12 public school system. She also teaches mathematics education for student teachers at UBC. For her master's research, she wrote about her experiences teaching Aboriginal art as a non-Aboriginal person.

Joel Grant is a member of the Métis Nation of Alberta (Region 3, Treaty 7 territory). He grew up playing ice hockey and with a passion for science. With support from his family, Joel was able to pursue a degree in engineering. He is passionate about giving back to Indigenous youth through initiatives such as the American Indian Science and Engineering Society and the Eagle Spirit Science Futures Camp. Joel will be doing a one-year Pathy Fellowship working on a project titled, "Indigenous Storytelling with Science Activities on Film."

Alex Allard-Gray hails from the Listuguj Mi'gmaq First Nation located in the Gaspé region of Quebec and works with the Indigenous Health Professions Program (IHPP) at McGill University. Alex's initial experience of post-secondary education came from participating in a McGill initiative that allowed Indigenous youth to be exposed to university life. Alex completed his B.Sc. in Physiology at McGill, where he became involved with Indigenous-led initiatives. This would include being a founder and later acting as president of the McGill Chapter of the Canadian Indigenous Science and Engineering Society (CaISES).

Marc Higgins is an Assistant Professor in the Department of Secondary Education at the University of Alberta and is affiliated with the Faculty of Education's Aboriginal Teacher Education Program (ATEP). As a white settler scholar, his research labours the methodological space within and between Indigenous, post-structural and post-humanist theories in order to (re)think and practice education which works to ethically respond to contested ways-of-knowing (i.e., epistemology) and ways-of-being (i.e., ontology) such as Indigenous science or ways-of-living-with-Nature. He recently published a book on this topic: *Unsettling Responsibility in Science Education—Indigenous Science, Deconstruction, and the Multicultural Science Education Debate.*

Eun-Ji Amy Kim is a first-generation settler from South Korea. A former high school teacher, she worked as a science curriculum consultant at Kahnawà:ke Education Center. Amy is a lecturer in Science Education at the School of Education and Professional Studies, Griffith University, Australia. She continues to work alongside with Indigenous peoples around the world. Amy also continues to grapple with the question of "how can citizens around the world live harmoniously? " She is now undertaking various projects on the topic of Global Citizenship Education by collaborating with individuals and organisations from different regions around the globe including Bangladesh, Canada, South Korea, the Philippines, USA and UK.

Kelly King is a settler with maternal ancestral roots in Poland and Latvia, and paternal ancestral roots in Scotland and England. Kelly has a B.A. in Indigenous Studies from Trent University (2014) and a Master's degree in Environmental Studies from York University (2017). Kelly is passionate about creating spaces to engage youth with traditional ecological knowledge and believes that by localizing our environmental perspectives, we can collectively make global differences. At the time of writing, Kelly was the Outreach Education Coordinator for TRACKS Youth Program.

Madison Laurin is a settler-Canadian with a family of British, French, Hungarian and German descent. Madison is a University of Toronto graduate with a Specialist B.A in Anthropology. Throughout her life, Madison has been involved in research and advocacy work with Indigenous groups in North and Central America; from Alberta to Hawai'i, Guatemala, Panama and Belize. She is excited to continue this work of learning from and connecting with communities while supporting land-based learning for Indigenous and non-Indigenous youth. At the time of writing, Madison was the Operations Coordinator for TRACKS Youth Program.

Shannon Leddy is a Vancouver based educator and writer. Shannon is a member of the Métis Nation with ties to the Red River Community in Manitoba from the 19th century and St. Louis, Saskatchewan from the 20th century. She was raised on Treaty Six Territory, Saskatoon, Saskatchewan.Her research focuses on inviting pre-service teachers into dialogue with contemporary Indigenous art in order to develop decolonial literacies that help them avoid reproducing colonial stereotypes and misrepresentation. She serves as an Assistant Professor (Teaching) in Indigenous Education in the Faculty of Education at the University of British Columbia. Leddy is the Co-Director of the Institute for Environmental Learning, a UNESCO Regional Centre of Excellence, and is a Research Fellow with the Institute for Public Education/BC.

Lorrie Miller holds a Ph.D. from UBC in Curriculum Studies: Art Education. Her teaching and writing explore pedagogical approaches that are woven with her passion for textile art, and *slow* pedagogy. Questions surrounding pedagogies of care in education drive her academic curiosity: what should education look like in fragile contexts during challenging times? At UBC, she teaches *Textile Design and Pedagogical Approaches: Art Education*, is the Associate Director at the Institute for Veterans Education & Transition, and has co-edited: Borderless Higher Education for Refugees: Lessons from the Dadaab Refugee Camps.

Kristin Muskratt is a Michi Saagiig Anishinaabe-kwe from Oshkiigmong (Curve Lake First Nation). Her Michi Saagiig Anishinaabe ancestral roots are both maternal and paternal. She also has paternal ancestral roots from Ireland and Scotland. Kristin is passionate about creating space for Indigenous youth to gain confidence and Indigenous pride by offering opportunities to learn more about traditional knowledge and fostering leadership skills. In her community, Kristin is involved with Youth for Water, Sacred Water Circle and the Trent Source Water Protection Committee. At the time of writing, Kristin was the Oshkwazin Coordinator for TRACKS Youth Program.

Ro'nikonhkátste Norton is a Kanien'kéha language learner, instructor, and advocate. He currently teaches Kanien'kehá:ka at Kanien'kéha Ratiwennahní:rats Adult Language Immersion Program.

Hannah Karahkwenhawe Stacey is a Mohawk woman from Kahnawake. She is a pre-service teacher enrolled in the B.Ed program at Kahnawake Education Center-McGill University. Her previous academic experiences include an Attestation of College Studies (AEC) in Early Childhood education from Champlain College St-Lambert and a Degree of College Studies (DEC) in Pure and Applied Science from Marianopolis College. In between classes, she tutors elementary and high school students in the community in Math, Science, English and French. In her spare time, she is usually found reading a fantasy or mystery novel.

Simon Sylliboy: Is from Eskasoni First Nation and he is a part-time instructor of education at St. Francis Xavier University. He is currently a Ph.D. student in educational studies focusing on decolonizing theories and Land-Based education. His background is in science education and has taught in his home community for 6 years. He is committed to revitalizing Mi'kmaq language and ways of knowing through Land-based education. He will be teaching Sociology of Education, curriculum and instruction in elementary science this upcoming Fall semester.

Dawn Wiseman is an Associate Professor in the School of Education at Bishop's University in Ktinékétolékouac (Sherbrooke, Québec, Canada). She has engaged in thinking about STEAM with young people and educators for over three decades, most often alongside Indigenous people, peoples, and communities in what is currently Canada. Her research exams how Western and Indigenous ways of knowing, being, and doing, might circulate together in STEAM education, student-directed STEAM inquiry, the distinctiveness of Canadian science education research, and the possibilities of teaching and learning within the context of human-driven climate change.

Bios-Mythois

Rehumanizing STEM through creative narratives
and humanizing approaches

Jennifer D. Adams
Series Editor

Bios-Mythois posits that we, as humans, are biological storytellers. Bios-Mythois is an innovative science education series that re-engages narratives, storytelling, and creative communicative structures in scientific ways of knowing and knowledge production. This series aims to refuse the Darwinian notion of being human that situates us as solely biological beings, but rather, reengages with the notion of using different modes of storytelling to develop collective and collaborative understandings of our world. Recognizing that language structures how we view and relate to nature and science, this series will expand the meanings of science in ways that include different accounts, communicative structures, and voices in ways that allow for nonlinear engagements and creative expansions of scientific understandings and pursuits. This series looks to re-engage the creative with the scientific; to re-engage humanity and more-than-humanity with lived narratives and texts about living, being, and becoming a part of our natural and built worlds. It will create a space within science, science education, and related studies that allow for expanded narratives about *what* science is, *who* can do science, and *how* science is communicated. We will focus on diverse voices, collaborative projects, and justice-oriented framings to advance new ways of scientific thought.

For additional information about this series or the submission of manuscripts, please contact:

Jennifer D. Adams Editor
jennifer.adams1@ucalgary.ca

To order other books in this series, please contact our Customer Service Department:

peterlang@presswarehouse.com (within the U.S.)
orders@peterlang.com (outside the U.S.)

Or browse online by series:

www.peterlang.com